RSV Epidemiological Surveillance

RSV Epidemiological Surveillance

Guest Editors

Emanuele Castagno
Irene Raffaldi

Basel • Beijing • Wuhan • Barcelona • Belgrade • Novi Sad • Cluj • Manchester

Guest Editors
Emanuele Castagno
Department of Pediatric Emergency
Regina Margherita Children's Hospital
Turin
Italy

Irene Raffaldi
Department of Pediatric Emergency
Regina Margherita Children's Hospital
Turin
Italy

Editorial Office
MDPI AG
Grosspeteranlage 5
4052 Basel, Switzerland

This is a reprint of the Special Issue, published open access by the journal *Viruses* (ISSN 1999-4915), freely accessible at: https://www.mdpi.com/journal/viruses/special_issues/5LSM5ND0X8.

For citation purposes, cite each article independently as indicated on the article page online and as indicated below:

Lastname, A.A.; Lastname, B.B. Article Title. *Journal Name* **Year**, *Volume Number*, Page Range.

ISBN 978-3-7258-3181-4 (Hbk)
ISBN 978-3-7258-3182-1 (PDF)
https://doi.org/10.3390/books978-3-7258-3182-1

Cover image courtesy of Emanuele Castagno

Contents

Preface

The new epidemiological trends in RSV are one of the most interesting and widely discussed topics at a global level, not only in the pediatric field.

The main aim of this Special Issue is to provide updated and high-quality evidence on the spread of RSV and its clinical manifestations in different contexts and in different geographical areas of the world. We are confident that the papers collected in this reprint may contribute to enhance the response of healthcare systems to the new challenges we must face.

This Special Issue has been produced thanks to the contribution of clinicians, virologists, and epidemiologists from different parts of the world. It is aimed not only at pediatric clinicians involved in the various phases of care of patients with symptomatic RSV infection, from primary care to emergency departments and intensive care units, but also at those involved in public health management.

We would like to thank all those who contributed to the realization of this Special Issue. First, we'd like to thank all the authors, reviewers, and editorial staff for their assistance. Last but not least, we warmly thank Mrs. Corrina Zhai for the invaluable assistance during all phases of the process.

Emanuele Castagno and Irene Raffaldi
Guest Editors

viruses

Editorial

The Epidemiology of Respiratory Syncytial Virus: New Trends and Future Perspectives

Irene Raffaldi and Emanuele Castagno *

Department of Pediatric Emergency, Regina Margherita Children's Hospital, 10126 Turin, Italy; iraffaldi@cittadellasalute.to.it
* Correspondence: ecastagno@cittadellasalute.to.it; Tel.: +390-113135205

Citation: Raffaldi, I.; Castagno, E. The Epidemiology of Respiratory Syncytial Virus: New Trends and Future Perspectives. *Viruses* **2024**, *16*, 1100. https://doi.org/10.3390/v16071100

Received: 2 July 2024
Accepted: 4 July 2024
Published: 8 July 2024

RSV (respiratory syncytial virus) is a major cause of acute lower respiratory tract infection (LRTI) worldwide. Immunocompromised adults and the elderly are susceptible to severe infection and high mortality. Among children, RSV is the most frequent cause of LRTI, particularly in infants and toddlers younger than 2 years old, and one half of this group will have been infected twice by this age [1,2].

Human RSV is an enveloped, spherical RNA virus belonging to the genus *Orthopneumovirus* in the family *Pneumoviridae* [1]; when infection develops, RSV forms large cells, known as syncytia. The structure of the virus consists of three membrane proteins: a small hydrophobic (SH) protein, an attachment glycoprotein (G), and a fusion (F) protein. Based on the protein G sequence, subtypes A and B can be identified. Both subtypes circulate simultaneously during an epidemic season, but usually one of the two predominates each year. The replication cycle of this virus is error-prone, allowing for a rapid generation of mutations, thus resulting in changes in the virulence of RSV and difficulties in creating antiviral agents or vaccines [3].

Its physiopathology consists of the disruption of the alveolar epithelium; this creates submucosal oedema and cilia loss on the apical surface of infected epithelial cells, together with an accumulation of mucus and cellular debris, inducing airway plugging and neutrophil infiltration in the airways. The severity of its clinical manifestations directly correlates with the virus titer and the exacerbated proinflammatory cytokine/chemokine response skewed toward a T helper type 2 immune response. However, the relative contributions of these factors to the determination of clinical features remain unknown [4].

Most children infected with RSV typically exhibit mild respiratory symptoms. The severe manifestations of RSV include pneumonia and bronchiolitis; the latter is usually self-limiting but accounts for a significant number of hospitalizations and admissions to pediatric intensive care units (PICUs), even in previously healthy, full-term newborns and infants [3].

Before the COVID-19 pandemic, RSV was typically described as a seasonal virus, characterized by a predictable epidemiological pattern, depending on the geographic area and climate [3]. In the Northern Hemisphere, the virus usually spread between November and March, with peak incidences in January/February, whereas in the Southern Hemisphere, RSV season typically occurred between June and September [3].

During the COVID-19 pandemic, drastic interventions were adopted on a global scale, and these included social distancing, stay-at-home orders, wearing face masks, and the promotion of hand washing. Such non-pharmaceutical interventions limited the circulation of SARS-CoV-2, together with other viruses transmitted through aerosol, droplets, and direct contact, such as RSV. In 2020, many countries saw an absence of RSV circulation during the "traditional" epidemiological season. For example, in Western Australia, a decrease of 98% in the detection of RSV infection was reported, compared to previous winter seasons between 2012 and 2019 [5]. Similarly, other authors reported a marked reduction in RSV spread in Europe and in other regions of the Northern Hemisphere, with a complete absence of cases noted in some countries [6–10].

After the initial reduction in RSV circulation, many authors recorded an out-of-season surge of RSV as the distancing measures adopted during the COVID-19 pandemic loosened. For example, in the Southern Hemisphere, a sharp rebound was observed in South Africa and New Zealand in late 2020 and early 2021, respectively [2]. Additionally, in the Northern Hemisphere, a similar rebound was also reported in Japan [11–13]; in the USA, where the total number of RSV infections had remained lower than expected until early 2021, a significant increase in reported infections was shown in statistics for the summer of 2021, largely before the traditional RSV outbreak period [11,12]. As regards Europe, the number of visits to the emergency department for acute bronchitis or bronchiolitis in England diverged from the seasonal trend, beginning in week 22 of 2021 and continuing onwards. In this survey, during the summer of 2021, the authors observed 9789 (84.9% [84.5 to 85.4]) more admissions than expected [11,14]. In addition to this, an Italian multicenter study reported a significantly higher admission rate for RSV infections in 2021 compared to 2020 and 2019 [3].

Epidemiological surveillance suggested that this new trend in the epidemic wave was not only present in children. During the second part of 2021, an inter-seasonal rise in RSV cases was observed even in younger adults, while an upsurge in older adults was observed later on [15].

Concerning pediatric patients in particular, some evidence for a correlation between affected patients of different ages and a higher severity of RSV-related clinical pictures arose after the first wave of the COVID-19 pandemic. On the one hand, Pruccoli et al. reported a shift towards older children presenting with symptomatic RSV infections [3]. Such an age shift was also confirmed by another report in Iceland; the median age of RSV-positive cases was 16 months during the 2020–2021 season, compared to 5.7 months across the five previous seasons [8,16].

On the other hand, Cai et al. recorded higher levels of severity in RSV-related disease and reported a percentage of 8.5% of RSV patients admitted to the PICU in 2022–2023, compared to 6.8% in pre-COVID-19 seasons. The authors also described an increased need for respiratory support than was previously reported (6.1% vs. 3.8%, respectively) [17]. However, such an observation has not been noticed by other authors, and this requires further confirmation in future epidemic seasons [3,18,19].

The COVID-19 pandemic scenario has offered a unique opportunity for researchers to learn more about the transmission of RSV and other respiratory viruses, as well as design future preventive strategies regarding RSV spread in both children and adults. Monitoring RSV seasonality is essential for pediatric health care in order to plan the appropriate timing of preventive strategies. The introduction of prophylactic measures, such as nirsevimab administration to newborns and infants, as well as the immunization of pregnant women, could be greatly influential in managing the next epidemic seasons from both clinical and organizational perspectives.

Author Contributions: I.R. and E.C. contributed equally to this editorial. All authors have read and agreed to the published version of the manuscript.

Funding: This paper received no external funding.

Conflicts of Interest: The authors declare no conflicts of interest.

References

1. Shang, Z.; Tan, S.; Ma, D. Respiratory syncytial virus: From pathogenesis to potential therapeutic strategies. *Int. J. Biol. Sci.* **2021**, *17*, 4073–4091. [CrossRef] [PubMed]
2. Eden, J.-S.; Sikazwe, C.; Xie, R.; Deng, Y.-M.; Sullivan, S.G.; Michie, A.; Levy, A.; Cutmore, E.; Blyth, C.C.; Britton, P.N.; et al. Off-season RSV epidemics in Australia after easing of COVID-19 restrictions. *Nat. Commun.* **2022**, *13*, 2884. [CrossRef] [PubMed]

3. Pruccoli, G.; Castagno, E.; Raffaldi, I.; Denina, M.; Barisone, E.; Baroero, L.; Timeus, F.; Rabbone, I.; Monzani, A.; Terragni, G.M.; et al. The Importance of RSV Epidemiological Surveillance: A Multicenter Observational Study of RSV Infection during the COVID-19 Pandemic. *Viruses* **2023**, *15*, 280. [CrossRef] [PubMed]
4. Cervantes-Ortiz, S.L.; Cuervo, N.Z.; Grandvaux, N. Respiratory Syncytial Virus and Cellular Stress Responses: Impact on Replication and Physiopathology. *Viruses* **2016**, *8*, 124. [CrossRef] [PubMed]
5. Yeoh, D.K.; Foley, D.A.; Minney-Smith, C.A.; Martin, A.C.; Mace, A.O.; Sikazwe, C.T. The impact of COVID-19 public health measures on detections of influenza and respiratory syncytial virus in children during the 2020 Australian winter. *Clin. Infect. Dis.* **2020**, *72*, 2199–2202. [CrossRef] [PubMed]
6. Groves, H.E.; Papenburg, J.; Mehta, K.; Bettinger, J.A.; Sadarangani, M.; Halperin, S.A.; Morris, S.K.; Bancej, C.; Burton, C.; Embree, J.; et al. The effect of the COVID-19 pandemic on influenza-related hospitalization, intensive care admission and mortality in children in Canada: A population-based study. *Lancet Reg. Health-Am.* **2022**, *7*, 100132. [CrossRef] [PubMed]
7. Lange, M.; Happle, C.; Hamel, J.; Dördelmann, M.; Bangert, M.; Kramer, R.; Eberhardt, F.; Panning, M.; Heep, A.; Hansen, G.; et al. Non-appearance of the RSV season 2020/21 during the COVID-19 pandemic–prospective, multicenter data on the incidence of respiratory syncytial virus (RSV) infection. *Dtsch. Arztebl. Int.* **2021**, *118*, 561–562. [PubMed]
8. Stein, R.T.; Zar, H.J. RSV through the COVID-19 pandemic: Burden, shifting epidemiology, and implications for the future. *Pediatr. Pulmonol.* **2023**, *58*, 1631–1639. [CrossRef] [PubMed]
9. Vittucci, A.C.; Piccioni, L.; Coltella, L.; Ciarlitto, C.; Antilici, L.; Bozzola, E.; Midulla, F.; Palma, P.; Perno, C.F.; Villani, A. The disappearance of respiratory viruses in children during the COVID-19 pandemic. *Int. J. Environ. Res. Public Health* **2021**, *18*, 9550. [CrossRef] [PubMed]
10. Van Brusselen, D.; De Troeyer, K.; Ter Haar, E.; Vander Auwera, A.; Poschet, K.; Van Nuijs, S.; Bael, A.; Stobbelaar, K.; Verhulst, S.; Van Herendael, B.; et al. Bronchiolitis in COVID-19 times: A nearly absent disease? *Eur. J. Pediatr.* **2021**, *180*, 1969–1973. [CrossRef] [PubMed]
11. Principi, N.; Autore, G.; Ramundo, G.; Esposito, S. Epidemiology of Respiratory Infections during the COVID-19 Pandemic. *Viruses* **2023**, *15*, 1160. [CrossRef] [PubMed]
12. Centers for Disease Control and Prevention. Changes in Influenza and Other Respiratory Virus Activity during the COVID-19 Pandemic—United States, 2020–2021. *Morb. Mortal. Wkly. Rep.* **2021**, *70*, 1013–1019. [CrossRef] [PubMed]
13. Ujiie, M.; Tsuzuki, S.; Nakamoto, T.; Iwamoto, N. Resurgence of Respiratory Syncytial Virus Infections during COVID-19 Pandemic, Tokyo, Japan. *Emerg. Infect. Dis.* **2021**, *27*, 2969–2970. [CrossRef]
14. Bardsley, M.; Morbey, R.A.; Hughes, H.E.; Beck, C.R.; Watson, C.H.; Zhao, H.; Ellis, J.; Smith, G.E.; Elliot, A.J. Epidemiology of respiratory syncytial virus in children younger than 5 years in England during the COVID-19 pandemic, measured by laboratory, clinical, and syndromic surveillance: A retrospective observational study. *Lancet Infect. Dis.* **2023**, *23*, 56–66. [CrossRef] [PubMed]
15. Honemann, M.; Thiem, S.; Bergs, S.; Berthold, T.; Propach, C.; Siekmeyer, M.; Frille, A.; Wallborn, T.; Maier, M.; Pietsch, C. In-Depth Analysis of the Re-Emergence of Respiratory Syncytial Virus at a Tertiary Care Hospital in Germany in the Summer of 2021 after the Alleviation of Non-Pharmaceutical Interventions Due to the SARS-CoV-2 Pandemic. *Viruses* **2023**, *15*, 877. [CrossRef] [PubMed]
16. van Summeren, J.; Meijer, A.; Aspelund, G.; Casalegno, J.S.; Erna, G.; Hoang, U.; Lina, B.; de Lusignan, S.; Teirlinck, A.C.; Thors, V.; et al. Low levels of respiratory syncytial virus activity in Europe during the 2020/21 season: What can we expect in the coming summer and autumn/winter? *Eurosurveillance* **2021**, *26*, 2100639. [CrossRef]
17. Cai, W.; Köndgen, S.; Tolksdorf, K.; Dürrwald, R.; Schuler, E.; Biere, B.; Schweiger, B.; Goerlitz, L.; Haas, W.; Wolff, T.; et al. Atypical age distribution and high disease severity in children with RSV infections during two irregular epidemic seasons throughout the COVID-19 pandemic, Germany, 2021 to 2023. *Eurosurveillance* **2024**, *29*, 2300465. [CrossRef] [PubMed]
18. Castagno, E.; Raffaldi, I.; Del Monte, F.; Garazzino, S.; Bondone, C. New epidemiological trends of respiratory syncytial virus bronchiolitis during COVID-19 pandemic. *World J. Pediatr.* **2023**, *19*, 502–504. [CrossRef]
19. Bourdeau, M.; Vadlamudi, N.K.; Bastien, N.; Embree, J.; Halperin, S.A.; Jadavji, T.; Langley, J.M.; Lebel, M.H.; Le Saux, N.; Moore, D.; et al. Pediatric RSV-Associated Hospitalizations before and during the COVID-19 Pandemic. *JAMA Netw. Open* **2023**, *6*, e2336863. [CrossRef] [PubMed]

viruses

MDPI

Article

Respiratory Syncytial Virus in Adult Patients at a Tertiary Care Hospital in Germany: Clinical Features and Molecular Epidemiology of the Fusion Protein in the Severe Respiratory Season of 2022/2023

Mario Hönemann [1,2,*], Melanie Maier [1,2], Armin Frille [3], Stephanie Thiem [1], Sandra Bergs [1], Thomas C. Williams [4], Vicente Mas [5], Christoph Lübbert [2,6] and Corinna Pietsch [1,2]

1 Virology Department, Institute of Medical Microbiology and Virology, Leipzig University Hospital, Johannisalle 30, 04103 Leipzig, Germany
2 Interdisciplinary Center for Infectious Diseases, Leipzig University Hospital, Liebigstrasse 20, 04103 Leipzig, Germany
3 Department of Respiratory Medicine, Leipzig University Hospital, Liebigstrasse 20, 04103 Leipzig, Germany
4 Child Life and Health, University of Edinburgh, Royal Hospital for Children and Young People, 50 Little France Crescent, Edinburgh EH16 4TJ, UK
5 Centro Nacional de Microbiología and CIBER de Enfermedades Respiratorias, Instituto de Salud Carlos III, C/ Sinesio Delgado, 4, 28029 Madrid, Spain
6 Division of Infectious Diseases and Tropical Medicine, Department of Medicine I, Leipzig University Hospital, Liebigstrasse 20, 04103 Leipzig, Germany
* Correspondence: mario.hoenemann@medizin.uni-leipzig.de

Citation: Hönemann, M.; Maier, M.; Frille, A.; Thiem, S.; Bergs, S.; Williams, T.C.; Mas, V.; Lübbert, C.; Pietsch, C. Respiratory Syncytial Virus in Adult Patients at a Tertiary Care Hospital in Germany: Clinical Features and Molecular Epidemiology of the Fusion Protein in the Severe Respiratory Season of 2022/2023. *Viruses* **2024**, *16*, 943. https://doi.org/10.3390/v16060943

Academic Editors: Emanuele Castagno and Irene Raffaldi

Received: 10 May 2024
Revised: 31 May 2024
Accepted: 2 June 2024
Published: 12 June 2024

Abstract: Following an interseasonal rise in mainly pediatric respiratory syncytial virus (RSV) cases in Germany in 2021, an exceptionally high number of adult cases was observed in the subsequent respiratory season of 2022/2023. The aim of this study was to compare the clinical presentation of RSV infections in the pre- and post-SARS-CoV-2 pandemic periods. Additionally, the local epidemiology of the RSV fusion protein was analyzed at a molecular genetic and amino acid level. RSV detections in adults peaked in calendar week 1 of 2023, 8 weeks earlier than the earliest peak observed in the three pre-pandemic seasons. Although the median age of the adult patients was not different (66.5 vs. 65 years), subtle differences between both periods regarding comorbidities and the clinical presentation of RSV cases were noted. High rates of comorbidities prevailed; however, significantly lower numbers of patients with a history of lung transplantation ($p = 0.009$), chronic kidney disease ($p = 0.013$), and immunosuppression ($p = 0.038$) were observed in the 2022/2023 season. In contrast, significantly more lower respiratory tract infections ($p < 0.001$), in particular in the form of pneumonia ($p = 0.015$) and exacerbations of obstructive lung diseases ($p = 0.008$), were detected. An ICU admission was noted for 23.7% of all patients throughout the study period. Sequence analysis of the fusion protein gene revealed a close phylogenetic relatedness, regardless of the season of origin. However, especially for RSV-B, an accumulation of amino acid point substitutions was noted, including in antigenic site Ø. The SARS-CoV-2 pandemic had a tremendous impact on the seasonality of RSV, and the introduction of new vaccination and immunization strategies against RSV warrants further epidemiologic studies of this important pathogen.

Keywords: RSV; molecular epidemiology; fusion protein; respiratory infections; respiratory viruses

1. Introduction

Respiratory syncytial virus (RSV) is one of the most widespread respiratory pathogens and affects all age groups across the population [1]. Besides mild upper respiratory tract infections (URTIs), RSV can especially cause severe lower respiratory tract infections (LRTIs) in infants and young children below the age of two years [2]. RSV infections in adults seem to predominantly occur in the form of mild URTIs [3]. However, severe disease courses with subsequent hospitalization and the development of LRTIs, i.e., in the form of pneumonia, are frequently observed especially in high-risk patient groups [4–9]. In addition to age, comorbidities may result in increased susceptibility to infections with a higher risk for severe disease progression. These include chronic diseases affecting the lungs (chronic obstructive pulmonary disease [COPD]), cardiovascular system, kidneys, liver, central nervous system (stroke), endocrine system (diabetes mellitus), and obesity, as well as other co-factors, such as immunosuppressive drugs (i.e., in the context of solid organ transplantation) [10–13].

RSV is a negative-stranded RNA virus belonging to the genus *Orthopneumovirus* in the family *Pneumoviridae*. In Europe, infections typically occur in late fall and the winter months with a peak in February [14]. Based on genetic and antigenic diversity [14,15], RSV can be subdivided into two subtypes, RSV-A and RSV-B. The virion contains a lipid envelope with embedded glycoproteins, most notably the fusion protein (F) and the attachment glycoprotein (G). The G protein initiates cell binding, while the F protein mediates fusion of the virion with host cells.

Recent proposals for revised genotype descriptions identified the G ectodomain as the lowest common denominator suitable for RSV genotyping [16–18]. However, only a narrow genotype spectrum was observed in recent epidemiological studies, as a global predominance of single genotypes of RSV-A and -B [4,16–19] has evolved and was even further accelerated by the contact restriction measures in the wake of the SARS-CoV-2 pandemic [20,21]. At the same time, due to recent approvals (i.e. FDA and EMA) of several RSV vaccines intended for pregnant women and adults older than 60 years of age and the new highly potent monoclonal antibody nirsevimab for the RSV prophylaxis of neonates [22–24], in addition to the monoclonal antibody palivizumab, the clinical importance and need for molecular surveillance of the fusion protein has vastly increased since 2023. The structure of the fusion protein is highly complex [25] and features a conformational change between a metastable prefusion conformation [26] when incorporated into virus particles and a stable postfusion conformation [27] after membrane binding and fusion. Antibody binding to the prefusion confirmation may be associated with a highly potent neutralization of RSV [28]. Six antigenic sites were identified on the fusion protein, of which three are present only on the prefusion confirmation (sites Ø (zero), II, and V) and three on both conformational states (sites I, II, and IV). Thus far, particular clinical importance can be ascribed to site II, which represents the binding site of palivizumab [29], and site Ø, the main target region for the recently approved vaccines and nirsevimab [30].

Following behavioral changes and the implementation of extensive non-pharmaceutical interventions (NPIs) [20] in 2020 due to the SARS-CoV-2 pandemic, which included nationwide lockdowns with the subsequent closing of schools, daycare centers, and cultural events, RSV activity was greatly reduced during the winter of 2020/2021. Following the easing of NPIs, an interseasonal rise in RSV cases, especially in children and younger adults, was observed in the summer and fall of 2021 [4], while an upsurge in older adults was observed with a delay of one year in 2022 and 2023.

The aim of this study was to characterize the clinical significance of RSV infections of adult patients between seasons 2017/2018, 2018/2019, and 2019/2020 before the SARS-CoV-2 pandemic and the 2022/2023 season. Additionally, the local molecular F gene epidemiology between 2017 and 2023 was analyzed.

2. Materials and Methods

2.1. Sample Collection and Clinical Data

From October 2017 to September 2023, 17,251 respiratory samples from 9997 adult ($\geq$18 years) in- and outpatients were collected and tested for viral respiratory infections at the University Hospital of Leipzig, Germany.

Samples included nasal and/or naso-oropharyngeal swabs (44.7%, n = 7708), throat rinsing fluids (30.1%, n = 5200), bronchoalveolar lavage fluids (15.2%, n = 2630), tracheal secretions (8.8%, n = 1523), and sputum samples (1.1%, n = 190). Testing was initiated at the discretion of the responsible physician. To avoid bias caused by follow-up samples, re-testing within six weeks after initial detection was excluded. For this retrospective observational study, data relating to underlying medical conditions, comorbidities, and clinical parameters from the day of RSV detection were retrieved from patient charts. In the case of missing clinical information, the designation "(n/total)" indicates the respective cases for the total amount of available data. A body temperature > 38.0 °C was categorized as fever. The classification of URTIs and LRTIs was carried out according to the International Statistical Classification of Diseases and Related Health Problems (ICD-10-WHO) [31] and the diagnoses and information listed in the patients' records. Patients with any of the following conditions were considered immunocompromised: receiving active chemotherapy for cancer, chronic neutropenia (<1500/µL for more than three months), receiving steroids or other immunomodulatory medications prior to visit/admission, aplasia of the thymus gland, or reported HIV infection. Furthermore, comorbidities were classified according to the age-adjusted Charlson comorbidity index (CCI) [32]. Asthma, COPD, and Asthma-COPD Overlap Syndrome (ACOS) were summarized as obstructive lung disease (OLD). Anticholinergics and β2-adrenergic agonists (i.e., ipratropium bromide and salbutamol) were classified as bronchodilators. Bacterial and fungal pathogens were cultivated with standard microbiological techniques and considered co-infections if detection from respiratory samples or blood was documented. Viral co-infections were assessed through a multiplex test for respiratory viruses (see below) and established routine laboratory protocols for CMV [33,34], HSV [35], and VZV [36]. SARS-CoV-2 was analyzed with a commercially available SARS-CoV-2 RT-PCR assay for the Alinity m analyzer (Abbott, Chicago, IL, USA) targeting the viral RdRp- and N-genes.

2.2. Saisonality, Detection Intervals, Peak Detection

A respiratory season was defined as starting on October 1st and ending on September 30th of the following year. Deviating from this definition, RSV cases between August 2021 and February 2022 were considered to belong to the 2021/2022 season. Additionally, cases from September 2022 were considered as belonging to the 2022/2023 season. For the sake of this study, the main detection interval, equaling the epidemic period, was defined as the period between the first of two subsequent weeks and the last of two subsequent weeks where at least two RSV cases were identified per week. The peak detection of RSV is equivalent to the week/weeks where the most RSV cases were identified.

2.3. Meteorological Data

The data for the mean monthly air temperatures (Leipzig/Halle; Station ID: 2932) were obtained from the Climate Data Centre (CDC) of the Deutscher Wetterdienst (DWD) [37].

2.4. Nucleic Acid (NA) Extraction and RSV Detection

Total NA was extracted from 200 µL of the respiratory samples using the DNA and Viral NA Small Volume Kit on a MagNA Pure 96 instrument (both Roche, Mannheim, Germany) according to the manufacturer's instructions. NAs were stored in aliquots at −80 °C until further use. The presence of genomes of common respiratory viruses, including influenza viruses A and B, RSV-A and -B, parainfluenza viruses 1 to 4, human coronaviruses (including 229E, NL63, OC43, and HKU1), human metapneumoviruses, adenoviruses, human bocaviruses, rhinoviruses, and enteroviruses, was assessed using a multiplex panel assay (NxTAG RPP, Luminex corporation, Austin, TX, USA) according to the manufacturer's instructions. Samples that reacted to either one or both RSV targets of the assay were further analyzed.

2.5. Phylogenetic Analysis of the RSV F Gene Sequence

The complete viral fusion protein gene (F gene) was amplified using a protocol optimized for contemporary RSV genotypes. The resulting amplicons spanned a fragment of approximately 2150 bp for both RSV-A and RSV-B after the final nested PCR step. Sequences were generated using either of the following two approaches. (1) Sanger Sequencing was performed using the BigDye Terminator Sequencing Kit v1.1 and an ABI 3500 Genetic Analyzer (both Applied Biosystems, Foster City, CA, USA). (2) Oxford Nanopore Sequencing was performed using the Rapid Barcoding Kit 96, R9.4.1 flow cells and a GridION Mk1 (all Oxford Nanopore Technologies Ltd., Oxford, UK). Reads were basecalled in real time with high accuracy and re-basecalled with super accuracy. Reads with a length above 600 nt were mapped to reference sequences KY654514 (RSV-A) and KY684758 (RSV-B). For a detailed description of the reaction conditions and primers, see Supplementary Tables S1–S7. The sequencing approach was applied to at least ten randomly selected samples of the four seasons before 2022/2023 that were already genotyped in the G gene [4]. For the 2022/2023 season, the sequencing approach was applied to all RSV-A cases and to randomly selected RSV-B cases, including all fatal cases. The sequences obtained were generated using Geneious Prime software, version 2023.2.1, and submitted to GenBank (accession numbers PP084746-PP084890 and PP354071-PP354073). Separate phylogenetic trees for RSV-A and RSV-B were constructed at the nucleotide level using MEGA software, version 7, based on the maximum likelihood method. Bootstrap analyses were performed with 1000 replicates [38]. For the analysis of the relatedness of clinical isolates of the current study, two phylogenetic trees for RSV-A and RSV-B were constructed, including full-length fusion protein gene consensus sequences derived from reference strains proposed by Goya et al. [16] for the genotype analysis of RSV (Supplementary Tables S8 and S9).

2.6. Amino Acid Analysis

The obtained fusion (F) protein sequences were translated in order to compare the analyzed strains on an amino acid (AA) level and to the translated consensus reference strains of genotypes GA2.3.5 and GB5.0.5a. The AA changes were mapped to fusion protein subunits, regions, and antigenic sites according to Mas et al. [25]. The AA residues considered for the antigenic sites are provided in Supplementary Table S10.

2.7. Statistical Analysis

Statistical analyses were performed using IBM SPSS Statistics for Windows, version 29.0 (IBM Corp., Armonk, NY, USA). Continuous values were expressed as medians (interquartile range (IQR)) and categorical data as frequencies (percentages). A Mann–Whitney U test was performed to compare continuous variables. A chi-square test or Fisher's exact test was performed for categorical variables. Brackets indicate parameters that were analyzed in the same contingency table. All tests were two tailed. A p-level of <0.05 was considered significant. For post hoc pairwise comparisons of column proportions, a Bonferroni correction for multiple comparisons was applied.

3. Results

3.1. RSV Detection and Seasonality

In total, the presence of RSV RNA was confirmed by RT-PCR in 477 samples originating from 343 unique cases. The absolute numbers of RSV cases detected between 2017 and 2023 stratified by subtype are shown in Figure 1. The main detection intervals for the seasons before 2022/2023 were as follows: week 5 to week 18 of 2018 (season 2017/2018), week 5 to week 17 of 2019 (season 2018/2019), week 4 to week 14 of 2020 (season 2019/2020), and week 37 to week 46 of 2021 (season 2021/2022). For these seasons, weeks with the highest incidence of new RSV cases were weeks 11 and 14 (seven cases each), week 9 (nine cases), weeks 9 and 11 (five cases each), and week 40 (six cases), respectively. For the 2022/2023 season, the main detection interval was observed between week 47 of 2022 (November) and week 13 of 2023 (March). The highest incidence of new cases within a single week was observed in week 1 of 2023 (18 cases). The 2020/2021 season was considered absent.

The peak RSV detections coincided with the lowest temperatures of the year (Figure 2), except for the 2021/2022 season. The peak positivity rates for seasons 2017/2018, 2018/2019, 2019/2020, 2021/2022, and 2022/2023 were 9.3%, 6.8%, 3.8%, 9.4%, and 9.1%, respectively.

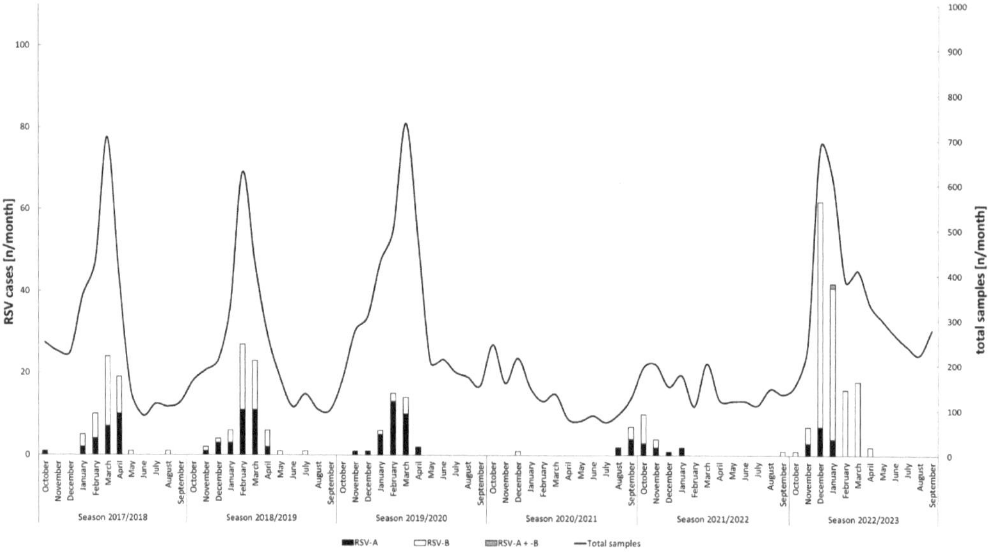

Figure 1. Monthly distribution of total number of samples tested and new RSV cases (n = 343) stratified by RSV-A and -B. Note the two different y-axes: the left y-axis refers to the bar charts and shows the absolute number of detected RSV cases while the right y-axis refers to the absolute number of samples tested, as represented by the line graph. The x-axis is labeled according to the definition that was used to define the start and end of a respiratory season.

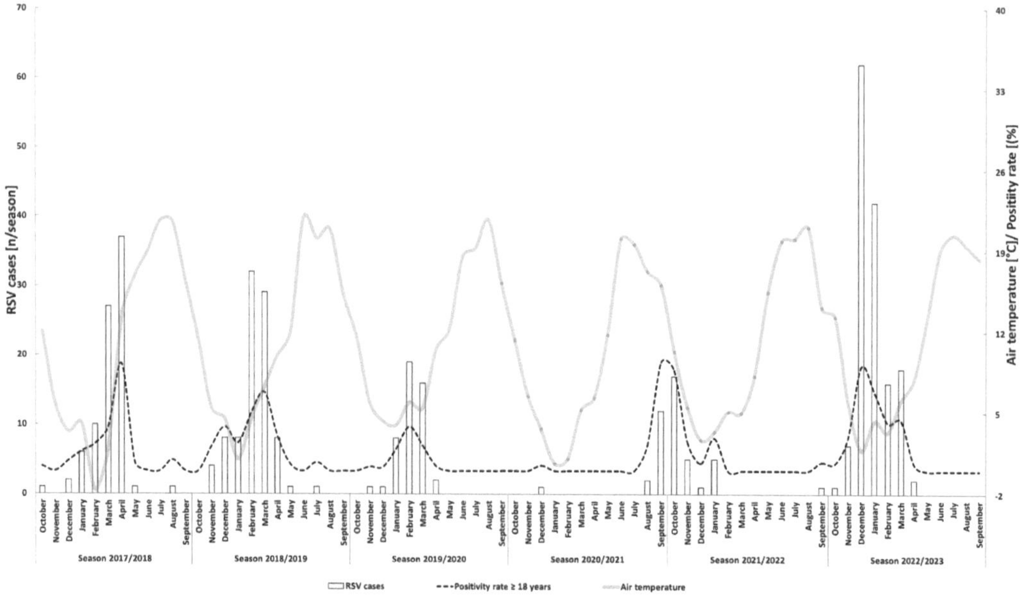

Figure 2. Monthly distribution of RSV cases (n = 343), positivity rates of the RSV assay, and mean air temperature during the study period. Note the two different y-axes: the left y-axis refers to the absolute numbers of detected new RSV cases (light gray bars) while the right y-axis refers to the mean temperature of the respective month [°C] (dark gray line) and the positivity rates for adult patients [%] (black dashed line). The x-axis is labeled according to the definition that was used to define the start and end of a respiratory season.

Two shifts between dominant RSV subtypes were observed during the study period ($p < 0.001$, Table 1). While most cases originated from infections with RSV-A in seasons 2019/2020 and 2021/2022, RSV-B infections were predominantly observed before and thereafter. The patients over the age of 60 predominated in the study population, except for the 2021/2022 season (Table 1).

Table 1. Sex, age, and RSV species distribution during the study period.

Seasons		2017/2018	2018/2019	2019/2020	2021/2022	2022/2023	Total	*p*-Value
Age group								
<60	[% (*n*/total)]	34.4 (21/61)	21.4 (15/70)	35.9 (14/39)	73.1 (19/26)	34.9 (51/146)	35.3 (121/343)	**<0.001**
≥60	[% (*n*/total)]	65. 6 (40/61)	78.6 (55/70)	64.1 (25/39)	26.9 (7/26)	65.1 (95/146)	64.7 (222/343)	
Sex								
male	[% (*n*/total)]	67.2 (41/61)	48.6 (34/70)	56.4 (22/39)	57.7 (15/26)	52.1 (76/146)	55.1 (189/343)	n.s.
female	[% (*n*/total)]	32.8 (20/61)	51.4 (36/70)	43.6 (17/39)	42.3 (11/26)	47.9 (70/146)	44.9 (154/343)	
RSV species								
RSV-A	[% (*n*/total)]	39.3 (24/61)	44.3 (31/70)	82.1 (32/39)	53.8 (14/26)	8.2 (12/146)	32.9 (113/343)	**<0.001**
RSV-B	[% (*n*/total)]	60.7 (37/61)	55.7 (39/70)	17.9 (7/39)	46.2 (12/26)	91.1 (133/146)	66.8 (229/343)	
mixed	[% (*n*/total)]	-	-	-	-	0.7 (1/146)	0.3 (1/343)	

Analyzed categories are displayed on the column to the left and given as relative and absolute frequencies [% (*n*/total)]. (*n*/total) indicates the respective cases of the total number of cases in the respective season. The *p*-values of the chi-square tests for the contingency tables, including all seasons and subgroups, are indicated. Significant *p*-values for the post hoc pairwise analysis of specific categories (age group) are given in the corresponding paragraph. n.s., not significant.

A relative age stratification into six age groups within RSV cases comparing the pre-pandemic seasons, the 2021/2022 season, and the 2022/2023 season is shown in Figure 3. A significant age difference was detected between those time periods ($p < 0.001$), originating from the high percentage of patients in the 25–44 years group in the 2021/2022 season in comparison to the pre-pandemic cohort ($p < 0.001$) and the patients of the 2022/2023 season ($p = 0.005$).

3.2. Study Population and Clinical Features

RSV-related patient characteristics and clinical parameters are presented in Table 2. A comparison between the pre-pandemic seasons (2017/2018, 2018/2019, and 2019/2020) and the 2022/2023 season was performed. Neither the median age nor the sex differed between these periods. The age-adjusted CCI of patients who contracted RSV was not statistically different between the two groups. However, a lower prevalence of patients with chronic kidney failure (26.4 vs. 39.9%, $p = 0.013$) and lung transplantation (0 vs. 4.7%, $p = 0.009$) and patients receiving immunosuppression (25.4 vs. 36.3, $p = 0.038$) were observed in the 2022/2023 season compared with RSV patients in the pre-pandemic seasons. Additionally, for the 2022/2023 season, fever ($p = 0.032$), LRTIs ($p < 0.001$), pneumonia ($p = 0.015$), exacerbation of OLD ($p = 0.008$), and administration of bronchodilators occurred more frequently compared to the pre-pandemic seasons. The detected co-pathogens are listed in Supplementary Table S11. A comparison between the 2021/2022 and 2022/2023 seasons is presented in Supplementary Table S12.

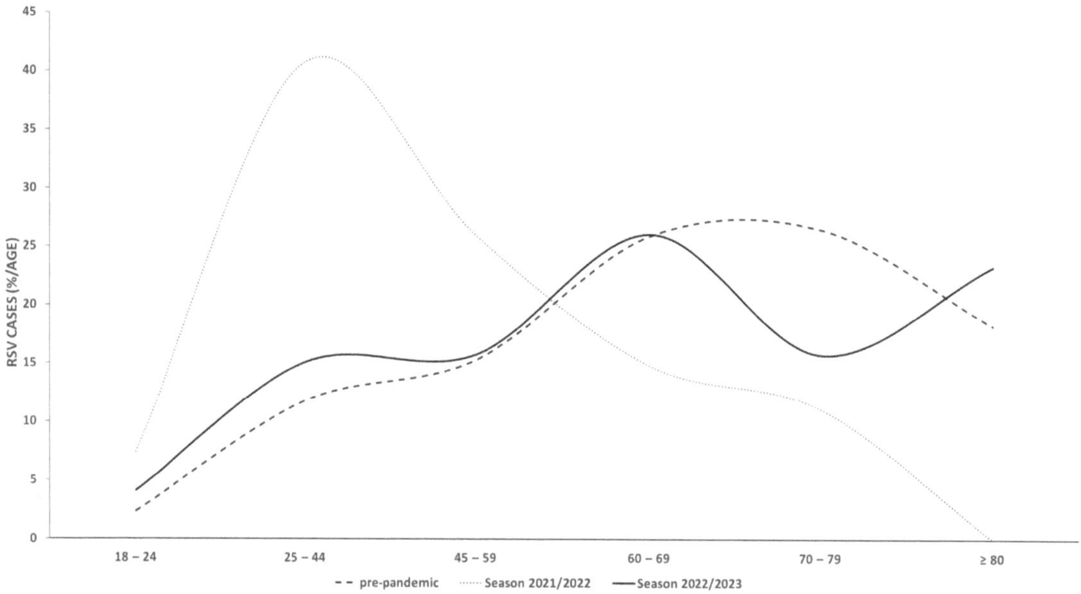

Figure 3. Relative distribution of RSV cases (n = 343) stratified by age. The lines represent the relative number of cases detected in the age group indicated. The 2017/2018, 2018/2019, and 2019/2020 seasons were designated "pre-pandemic".

3.3. Phylogenetic Analysis

For RSV-A, the F gene amplification approach was performed for 12, 12, 12, 11, and nine cases for the seasons from 2017/2018 to 2022/2023, respectively. For RSV-B, the gene amplification approach was performed for 11, 10, 10, 12, and 48 cases for the seasons from 2017/2018 to 2022/2023, respectively. Within the genotyped subset, respiratory specimens included 111 nasal and/or oropharyngeal swabs (75.5%), 20 throat rinsing fluids (13.6%), 14 bronchoalveolar lavage fluids (9.5%), and two tracheal secretions (1.4%). For all isolates, the coding sequence (CDS) had a length of 1,725 bases, translating into 574 amino acids.

The phylogenetic analysis of the complete F gene revealed a close relationship between all RSV-A isolates analyzed (Figure 4). The pairwise nucleotide identity was 99.5% for the isolates of the 2022/2023 season and 99% for all analyzed sequences. No distinct well-supported cluster was observed for any of the five seasons, as most groupings of different isolates are located in terminal nodes with low statistical support of ancestral nodes. The isolates clustered together with the consensus sequence of GA2.3.5 (Goya et al. [16]), which was also confirmed by a separate phylogenetic analysis carried out with proposed reference strains (Supplementary Figure S1). No distinct cluster of fatal cases was observed.

The phylogenetic analysis of the complete F gene revealed a close relationship between all analyzed RSV-B isolates (Figure 5). The pairwise nucleotide identity was 99.4% for the isolates of the 2021/2022 season and 99.3% for all analyzed sequences. The isolates clustered close to the consensus sequence of GB5.0.5a (Goya et al. [16]), which was also confirmed in the separate phylogenetic analysis with proposed reference strains (Supplementary Figure S2). Except for two strains in the 2017/2018 season, the majority of the isolates formed a distinct phylogenetic cluster that corresponded to distinct AA changes with regard to the GA2.3.5 consensus translation (see below). No distinct cluster of fatal cases was observed.

Table 2. Study population and clinical features of the 2022/2023 season and pre-pandemic RSV cases.

		Pre-Pandemic	2022/2023	Total	*p*-Value
Study population					
Female	[% (*n*/total)]	42.9 (73/170)	47.9 (70/146)	45.3 (143/316)	n.s.
Male	[% (*n*/total)]	57.1 (97/170)	52.1 (76/146)	54.7 (173/316)	
Age [years]	[median (IQR)] SD]	66.5 (56–77)	65 (49.75–78)	66 (55–78)	n.s.
Inpatients	[% (*n*/total)]	79.4 (135/170)	82.8 (120/145)	81 (255/315)	n.s.
Outpatients	[% (*n*/total)]	20.6 (35/170)	17.2 (25/145)	19 (60/315)	
Length of hospital stay [days]	[median (IQR)]	13 (7–24.5)	10 (4–20)	12 (6–22)	n.s.
Comorbidities and risk factors					
Obstructive lung disease [OLD]	[% (*n*/total)]	22.6 (38/168)	31.2 (44/141)	26.5 (82/309)	n.s.
Lung transplant	[% (*n*/total)]	4.7 (8/170)	0 (0/142)	2.6 (8/309)	**0.009**
Chronic kidney failure	[% (*n*/total)]	39.9 (67/159)	26.4 (37/140)	33.8 (104/308)	**0.013**
Heart failure	[% (*n*/total)]	19.0 (32/168)	17.0 (24/141)	18.1 (56/309)	n.s.
Arterial hypertension	[% (*n*/total)]	58.9 (99/168)	57 (81/142)	58.1 (180/310)	n.s.
Coronary heart disease	[% (*n*/total)]	16.7 (28/168)	14.9 (21/141)	15.9 (49/309)	n.s.
Diabetes	[% (*n*/total)]	31.5 (53/168)	24.8 (35/141)	28.5 (88/309)	n.s.
Immunosuppression	[% (*n*/total)]	36.3 (61/168)	25.4 (36/142)	31.3 (97/310)	**0.038**
Malignancy	[% (*n*/total)]	35.7 (60/168)	28.2 (40/142)	32.3 (100/310)	n.s.
Solid	[% (*n*/total)]	11.9 (20/168)	4.9 (7(142)	8.7 (27/310)	n.s.
Hematologic	[% (*n*/total)]	23.2 (39/168)	22.5 (32/142)	22.9 (71/310)	
Solid and hematologic	[% (*n*/total)]	0.6 (1/168)	0.7 (1/142)	0.6 (2/310)	
CCI	[median (IQR)]	5 (4–7)	5 (3–7)	5 (3–7)	n.s.
Clinical presentation and features					
Fever	[% (*n*/total)]	24.1 (38/158)	35.8 (43/120)	29.1 (81/178)	**0.032**
Newly reported dyspnea	[% (*n*/total)]	43.5 (67/154)	48.4 (59/122)	45.7 (126/276)	n.s.
URTI	[% (*n*/total)]	35.4 (28/79)	34.0 (33/97)	34.7 (61/176)	n.s.
LRTI	[% (*n*/total)]	52.5 (83/158)	77.8 (84/108)	62.8 (167/266)	**<0.001**
Bronchitis	[% (*n*/total)]	7.6 (12/158)	10.3 (11/107)	8.7 (23/265)	n.s.
Pneumonia	[% (*n*/total)]	33.3 (53/159)	48.1 (52/108)	39.3 (105/267)	**0.015**
Exacerbation of OLD	[% (*n*/total)]	13.8 (22/159)	26.9 (29/108)	19.1 (51/167)	**0.008**
ICU stay	[% (*n*/total)]	25.9 (44/170)	21.1 (30/142)	23.7 (74/312)	n.s.
Length of ICU stay [days]	[median (IQR)]	3 (1–10)	5 (2.75–12.5)	3.5 (2–10)	n.s.
Ventilatory support	[% (*n*/total)]	15.3 (26/170)	23.2 (33/142)	18.9 (59/312)	n.s.
None *	[% (*n*/total)]	84.7 (144/170)	76.8 (109/142)	81.1 (253/312)	n.s.
HFNC	[% (*n*/total)]	0.6 (1/170)	1.4 (2/142)	1.0 (2/312)	
Non-invasive	[% (*n*/total)]	7.1 (12/170)	9.2 (13/142)	8.0 (25/312)	
Invasive	[% (*n*/total)]	7.6 (13/170)	12.7 (18/142)	9.9 (31/312)	
Administration of bronchodilators	[% (*n*/total)]	16.9 (28/166)	30.9 (42/136)	23.2 (70/302)	**0.004**
Syst. prednisolone administration	[% (*n*/total)]	14.9 (25/168)	22.0 (29/132)	18.0 (54/300)	n.s.
Co-infections	[% (*n*/total)]	21.2 (36/170)	27.3 (39/143)	24.0 (75/313)	n.s.
Bacterial	[% (*n*/total)]	5.9 (10/170)	11.9 (17/143)	8.6 (27/313)	n.s.
Viral	[% (*n*/total)]	11.2 (19/170)	9.8 (14/143)	10.5 (33/313)	
Fungal	[% (*n*/total)]	1.8 (3/170)	0.7 (1/143)	1.3 (4/313)	
Combined	[% (*n*/total)]	2.4 (4/170)	5.6 (8/143)	3.8 (12/313)	
Mortality	[% (*n*/total)]	6.5 (11/170)	12 (17/142)	9 (28/312)	n.s.

Analyzed categories are displayed in the column to the left and are either given as frequencies (%) or as the median and interquartile range (median (IQR)). (n/total) indicates the respective cases for the total amount of available data. The *p*-values of the chi-square tests for the contingency tables, including all subcategories, are indicated. All significant *p*-values for the post hoc pairwise analysis are given in the corresponding paragraph. The Mann–Whitney U test was performed to compare continuous variables. CCI, Charlson comorbidity index; HFNC, high-flow nasal cannula; ICU, intensive care unit; LRTI, lower respiratory tract infection; n.s., not significant; OLD, obstructive lung disease; syst., systemic; URTI, upper respiratory tract infection; * including low-flow oxygen via nasal cannula.

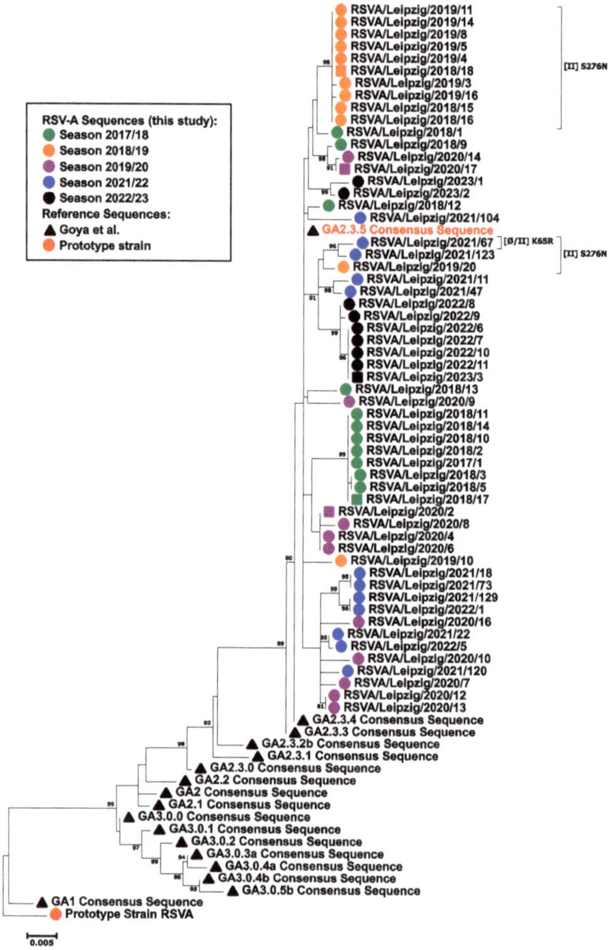

Figure 4. Molecular phylogenetic analysis of the RSV-A F gene by the maximum likelihood method. The evolutionary history was inferred using the maximum likelihood method based on the Tamura–Nei model [39]. The tree with the highest log likelihood (−4744.71) is shown. The percentage of trees in which the associated taxa clustered together is shown next to the branches. Initial tree(s) for the heuristic search were obtained automatically by applying Neighbor-Joining and BioNJ algorithms to a matrix of pairwise distances estimated using the Maximum Composite Likelihood (MCL) approach and then selecting the topology with superior log likelihood value. The tree is drawn to scale, with branch lengths measured in the number of substitutions per site. The analysis involved 75 nucleotide sequences. There were a total of 1725 positions in the final dataset. Evolutionary analyses were conducted in MEGA7 [38]. Only nodes with statistical support > 80% are shown. The following symbols indicate the sequence origin or the season of the indicated strain: dots/squares. Red, RSV-A prototype strain; green: season 2017/2018 isolates; orange, season 2018/2019 isolates; purple, season 2019/2020 isolates; blue, season 2021/2022 isolates; black, season 2022/2023 isolates; squares, fatal cases; black triangle: consensus reference sequences according to Goya et al. [16]. Amino acid changes in sites Ø and II with regard to the GA2.3.5 consensus sequence (red) are given in brackets.

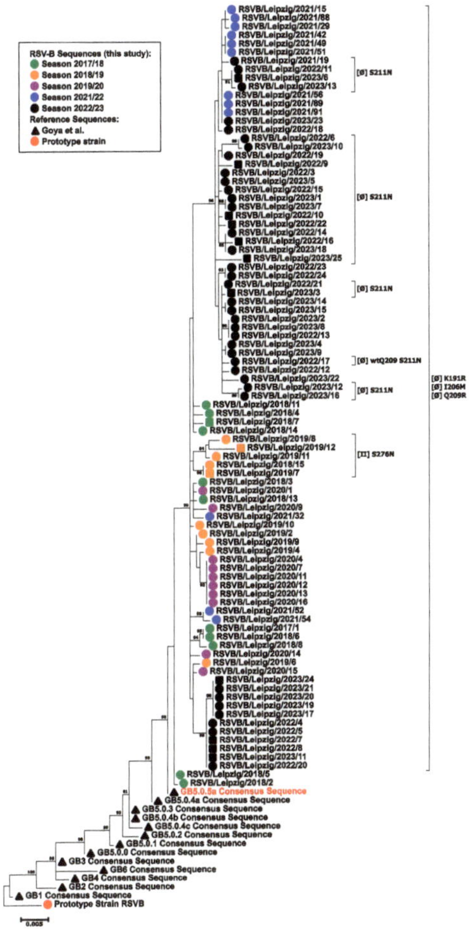

Figure 5. Molecular phylogenetic analysis of the RSV-B F gene by the maximum likelihood method. The evolutionary history was inferred using the maximum likelihood method based on the Tamura–Nei model [39]. The tree with the highest log likelihood (−4220.58) is shown. The percentage of trees in which the associated taxa clustered together is shown next to the branches. Initial tree(s) for the heuristic search were obtained automatically by applying Neighbor-Joining and BioNJ algorithms to a matrix of pairwise distances estimated using the Maximum Composite Likelihood (MCL) approach and then selecting the topology with superior log likelihood value. The tree is drawn to scale, with branch lengths measured in the number of substitutions per site. The analysis involved 105 nucleotide sequences. There were a total of 1725 positions in the final dataset. Evolutionary analyses were conducted in MEGA7 [38]. Only nodes with statistical support > 80% are shown. The following symbols indicate the sequence origin or the season of the indicated strain: dots/squares. Red, RSV-B prototype strain; green: season 2017/2018 isolates; orange, season 2018/2019 isolates; purple, season 2019/2020 isolates; blue, season 2021/2022 isolates; black, season 2022/2023 isolates; squares, fatal cases; black triangle: consensus reference sequences according to Goya et al. [16]. Amino acid changes in sites Ø and II with regard to the GA2.3.5 consensus sequence (red) are given in brackets. Note that for the RSVB/Leipzig/2022/17 strain, no amino acid change was noted at position 209 (wtQ209).

3.4. Amino Acid Analysis

The detected AA changes within the fusion protein are depicted in Tables 3 and 4 (F0 numbering). In comparison to the GA2.3.5 consensus translation, which was used as a reference, no distinct AA exchange was noted that was present in all isolates of the current study. The following AA changes were found in more than five percent of the strains: S276N (22.8%), T12I (22.1%), L20F (15.8%), S105N (14%), A10V (8.8%), and S377N (7%). The majority of AA changes were noted in the F1 subunit (56.7%), especially the signal peptide (41.8%). One AA exchange could be mapped to site Ø, K65R (1.8%), while site II was affected with both K65R and S276N substitutions. In comparison to the GB5.0.5a consensus translation, which was used as a reference, nearly all strains showed three distinct AA changes at positions 191, 206, and 209. In detail, the following substitutions were found in more than five percent of the strains: K191R, I206M (both 97.8%), Q209R (96.7%), S190N (51.7%), S211N, S389P (both 50.6%), K123R (12.1%), and S276N (5.5%). The majority of AA changes were noted in the F2 subunit (93.4%). Five AA changes (I64V, I206M, Q209R, S211N, and E295D) could be mapped to site Ø, affecting 97.8% of all strains, while substitutions in site II were observed in 6.6% of all strains (L273I and S276N). AA changes S190N, S211N, and S389P were highly frequent and only observed in strains from the 2021/22 season onwards.

Table 3. Amino acid changes in RSV-A isolates.

AA position	9	10	12	13	20	65	103	105	113	114	122	127	276	334	377	384	543	547
Subunit	F1	F1	F1	F1	F1	F1	F1	F1					F2	F2	F2	F2	F2	F2
Region	SP	SP	SP	SP	SP				p27	p27	p27	P27					CT	CT
Antigenic sites						Ø/II							II/III		I/III	I		
GA2.3.5 consensus	**N**	**A**	**T**	**T**	**L**	**K**	**A**	**S**	**R**	**F**	**T**	**V**	**S**	**L**	**S**	**I**	**A**	**L**
16 strains *	.	.	.	.	.	.	.	.	.	.	.	.	.	.	.	.	.	.
RSVA/Leipzig/2017/1	.	.	.	.	.	.	.	N	.	.	.	.	.	.	.	.	.	.
RSVA/Leipzig/2018/2	.	.	.	.	.	.	.	N	.	.	.	.	.	.	.	.	.	.
RSVA/Leipzig/2018/3	.	.	.	.	.	.	.	N	.	.	.	.	.	.	.	.	.	I
RSVA/Leipzig/2018/5	.	.	.	.	.	.	.	N	.	.	.	.	.	.	.	.	.	I
RSVA/Leipzig/2018/9	.	.	.	.	.	.	.	.	.	.	.	.	.	.	.	T	.	.
RSVA/Leipzig/2018/10	.	.	.	.	.	.	.	N	.	.	.	.	.	.	.	.	.	.
RSVA/Leipzig/2018/11	.	.	.	.	.	.	.	N	.	.	.	.	.	.	.	.	.	.
RSVA/Leipzig/2018/12	.	.	.	.	.	.	.	.	.	.	.	A	.	.	.	.	.	.
RSVA/Leipzig/2018/14	.	.	.	.	.	.	.	N	.	.	.	.	.	.	.	.	.	.
RSVA/Leipzig/2018/17	.	.	.	.	.	.	.	N	.	.	.	.	.	.	.	.	.	.
RSVA/Leipzig/2018/15	.	.	.	.	.	.	.	.	.	.	.	.	N	.	.	.	.	.
RSVA/Leipzig/2018/18	.	.	.	.	.	.	.	.	.	.	.	.	N	.	.	.	.	.
RSVA/Leipzig/2018/16	.	.	.	.	.	.	.	.	.	.	.	.	N	.	.	.	.	.
RSVA/Leipzig/2019/3	.	.	.	.	.	.	.	.	.	.	.	.	N	.	.	.	.	.
RSVA/Leipzig/2019/4	.	.	.	.	.	.	.	.	.	.	.	.	N	.	.	.	.	.
RSVA/Leipzig/2019/5	.	.	.	.	.	.	.	.	.	.	.	.	N	.	.	.	.	.
RSVA/Leipzig/2019/8	.	.	.	.	.	.	.	.	.	.	.	.	N	.	.	.	.	.
RSVA/Leipzig/2019/10	.	.	.	.	.	.	.	.	.	.	.	.	.	.	.	.	T	.
RSVA/Leipzig/2019/11	.	.	.	.	.	.	.	.	.	.	.	.	N	.	.	.	.	.
RSVA/Leipzig/2019/14	.	.	.	.	.	.	.	.	.	.	.	.	N	.	.	.	.	.
RSVA/Leipzig/2019/20	.	.	I	.	.	.	.	.	.	.	.	.	N	.	.	.	.	.
RSVA/Leipzig/2019/16	.	.	.	.	.	.	.	.	.	.	.	.	N	.	.	.	.	.
RSVA/Leipzig/2020/10	.	.	.	.	.	.	.	.	.	S	.	.	.	I	.	.	.	.
RSVA/Leipzig/2020/16	D	.	.	A	.	.	.	.	.	.	.	.	.	.	.	.	.	.
RSVA/Leipzig/2021/11	.	.	I	.	.	.	.	.	.	.	.	.	.	.	.	.	.	.
RSVA/Leipzig/2021/18	.	.	.	.	.	.	.	.	.	.	.	.	.	.	N	.	.	.
RSVA/Leipzig/2021/47	.	.	I	.	.	.	.	.	.	.	.	.	.	.	.	.	.	.
RSVA/Leipzig/2021/67	.	.	I	.	F	R	.	.	.	.	.	.	N	.	.	.	.	.
RSVA/Leipzig/2021/73	.	.	.	.	.	.	.	.	.	.	.	.	.	.	N	.	.	.
RSVA/Leipzig/2021/123	.	.	I	.	F	.	.	.	.	.	.	.	N	.	.	.	.	.
RSVA/Leipzig/2021/129	.	.	.	.	.	.	.	.	I	.	.	.	.	.	N	.	.	.
RSVA/Leipzig/2022/1	.	.	.	.	.	.	.	.	I	.	.	.	.	.	N	.	.	.
RSVA/Leipzig/2022/6	.	V	I	.	F	.	.	.	.	.	.	.	.	.	.	.	.	.
RSVA/Leipzig/2022/7	.	V	I	.	F	.	.	.	.	.	.	.	.	.	.	.	.	.
RSVA/Leipzig/2022/8	.	.	I	.	F	.	.	.	.	.	.	.	.	.	.	.	.	.
RSVA/Leipzig/2022/9	.	.	I	.	F	.	.	.	.	.	.	.	.	.	.	.	.	.
RSVA/Leipzig/2022/10	.	V	I	.	F	.	.	.	.	.	.	.	.	.	.	.	.	.
RSVA/Leipzig/2022/11	.	V	I	.	F	.	.	.	.	.	.	.	.	.	.	.	.	.
RSVA/Leipzig/2022/1	.	.	.	.	.	.	T	.	.	.	A	.	.	.	.	.	.	.
RSVA/Leipzig/2022/2	.	.	.	.	.	.	T	.	.	.	A	.	.	.	.	.	.	.
RSVA/Leipzig/2022/3	.	V	I	.	F	.	.	.	.	.	.	.	.	.	.	.	.	.

Amino acids (AA) are numbered with regard to the F0 protein. Additionally, the F1 and F2 subunits are indicated. The consensus amino acid residues to which the isolates were compared were derived from the GA2.3.5 consensus sequence. The AA changes were mapped to the following regions or antigenic sites: SP, signal peptide; p27, 27 amino acid fragments; CT, cytoplasmatic tail; Ø, site Ø; I–III, site I to III. * Sixteen strains that were identical at the amino acid level are depicted as a pool and contain isolates of the following seasons: two strains for 2017/2018, ten strains for 2019/2020, four strains for 2021/22.

Table 4. Amino acid changes in RSV-B isolates.

AA position	9	12	22	42	64	108	113	116	123	157	190	191	206	209	211	273	276	295	389	463	477	507	508	514	527
Subunit	F1	F1	F1	F1	F1	F1				F2	F2	F2	F2	F2	F2	F2	F2	F2	F2	F2	F2	F2	F2	F2	F2
Region	SP	SP	SP				p27	p27	p27	A	A	A										B	B	B	CT
Antigenic sites				I	Ø							V	Ø	Ø	Ø	II	I/II	Ø	I						
GB5.0.5a consensus	**S**	**F**	**L**	**R**	**I**	**R**	**Q**	**N**	**K**	**V**	**S**	**K**	**I**	**Q**	**S**	**L**	**S**	**E**	**S**	**E**	**Y**	**R**	**R**	**H**	**I**
2 strains [a]	.	.	.	.	.	.	.	.	.	.	.	.	.	.	.	.	.	.	.	.	.	.	.	.	.
25 strains [b]	.	.	.	.	.	.	.	.	.	.	.	R	M	R	.	.	.	.	.	.	.	.	.	.	.
RSVB/Leipzig/2018/3	.	.	.	.	.	.	.	.	.	.	.	R	M	R	.	.	.	D	.	.	.	.	.	.	.
RSVB/Leipzig/2018/13	.	.	.	.	.	.	.	.	.	.	.	R	M	R	.	.	.	.	.	.	.	.	.	.	V
RSVB/Leipzig/2018/15	.	.	.	.	.	.	.	.	.	.	.	R	M	R	.	.	N	.	.	.	.	.	.	.	.
RSVB/Leipzig/2019/7	.	.	.	.	.	.	.	.	.	.	.	R	M	R	.	.	N	.	.	.	.	.	.	.	.
RSVB/Leipzig/2019/8	.	L	.	.	.	.	.	.	.	.	.	R	M	R	.	.	N	.	.	D	.	.	.	.	.
RSVB/Leipzig/2019/11	.	L	.	.	.	.	.	.	.	.	.	R	M	R	.	.	N	.	.	.	.	.	.	.	.
RSVB/Leipzig/2019/12	.	L	F	.	.	.	.	.	.	.	N	R	M	R	.	.	N	.	.	D	.	.	.	.	.
RSVB/Leipzig/2021/19	.	.	.	.	.	.	.	.	.	A	N	R	M	R	N	.	.	.	P	.	.	.	.	.	.
21 strains [c]	.	.	.	.	.	.	.	.	.	.	N	R	M	R	N	.	.	.	P	.	.	.	.	.	.
RSVB/Leipzig/2022/3	.	.	.	K	.	.	.	.	.	.	N	R	M	R	N	.	.	.	P	.	.	.	.	.	.
RSVB/Leipzig/2022/4	.	.	.	.	.	.	.	.	R	.	.	R	M	R	.	.	.	.	.	.	.	.	.	.	.
RSVB/Leipzig/2022/5	.	.	.	.	.	.	.	.	R	.	.	R	M	R	.	.	.	.	.	.	.	.	.	.	.
RSVB/Leipzig/2022/6	.	.	.	.	.	.	.	.	.	.	N	R	M	R	N	.	.	.	P	.	.	H	.	R	.
RSVB/Leipzig/2022/7	.	.	.	.	.	.	.	.	R	.	.	R	M	R	.	.	.	.	.	.	.	.	.	.	.
RSVB/Leipzig/2022/8	.	.	.	.	.	.	.	.	R	.	.	R	M	R	.	.	.	.	.	.	.	.	.	.	.
RSVB/Leipzig/2022/9	.	.	.	K	.	.	.	.	.	.	N	R	M	R	N	.	.	.	P	.	.	.	.	.	.
RSVB/Leipzig/2022/10	.	.	.	K	V	.	.	.	.	.	N	R	M	R	N	.	.	.	P	.	.	.	.	.	.
RSVB/Leipzig/2022/11	.	.	.	.	.	.	.	S	.	.	N	R	M	R	N	.	.	.	P	.	.	.	.	.	.
RSVB/Leipzig/2022/14	.	.	.	K	.	.	.	.	.	.	N	R	M	R	N	.	.	.	P	.	.	.	.	.	.
RSVB/Leipzig/2022/15	.	.	.	K	.	.	.	.	.	.	N	R	M	R	N	.	.	.	P	.	.	.	.	.	.
RSVB/Leipzig/2022/16	N	.	.	K	.	.	.	.	.	.	N	R	M	R	N	.	.	.	P	.	.	.	.	.	.
RSVB/Leipzig/2022/17	.	.	.	.	.	.	.	.	.	.	N	R	M	.	N	.	.	.	P	.	.	.	.	.	.
RSVB/Leipzig/2022/18	.	.	.	.	.	.	R	.	.	.	N	R	M	R	N	.	.	.	P	.	.	.	.	.	.
RSVB/Leipzig/2022/19	.	.	.	K	.	.	.	.	.	.	N	R	M	R	N	.	.	.	P	.	.	.	.	.	.
RSVB/Leipzig/2022/20	.	.	.	.	.	.	.	.	R	.	.	R	M	R	.	.	.	.	.	.	.	.	.	.	.
RSVB/Leipzig/2022/21	.	.	.	.	.	.	.	.	.	.	N	R	M	R	N	.	.	.	P	.	.	.	.	.	.
RSVB/Leipzig/2022/22	.	.	.	K	.	K	.	.	.	.	N	R	M	R	N	.	.	.	P	.	.	.	.	.	.
RSVB/Leipzig/2023/1	.	.	.	K	.	.	.	.	.	.	N	R	M	R	N	.	.	.	P	.	.	.	.	.	.
RSVB/Leipzig/2023/3	.	.	.	.	.	.	.	.	.	.	N	R	M	R	N	I	.	.	P	.	.	.	.	.	.
RSVB/Leipzig/2023/5	.	.	.	K	.	.	.	.	.	.	N	R	M	R	N	.	.	.	P	.	.	.	.	.	.
RSVB/Leipzig/2023/6	.	.	.	.	.	.	.	S	.	.	N	R	M	R	N	.	.	.	P	.	.	.	.	.	.
RSVB/Leipzig/2023/7	.	.	.	K	.	.	.	.	.	.	N	R	M	R	N	.	.	.	P	.	.	.	.	.	.
RSVB/Leipzig/2023/10	.	.	.	.	.	.	.	.	.	.	N	R	M	R	N	.	.	.	P	.	.	H	.	R	.
RSVB/Leipzig/2023/11	.	.	.	.	.	.	.	.	R	.	.	R	M	R	.	.	.	.	.	.	.	.	.	.	.
RSVB/Leipzig/2023/12	.	.	.	.	.	.	.	.	.	.	N	R	M	R	N	.	.	.	P	.	H	.	.	.	.
RSVB/Leipzig/2023/13	.	.	.	.	.	.	.	S	.	.	N	R	M	R	N	.	.	.	P	.	.	.	.	.	.
RSVB/Leipzig/2023/16	.	.	.	.	.	.	.	.	.	.	N	R	M	R	N	.	.	.	P	.	H	.	.	.	.
RSVB/Leipzig/2023/17	.	.	.	.	.	.	.	.	R	.	.	R	M	R	.	.	.	.	.	.	.	.	.	.	.
RSVB/Leipzig/2023/18	.	.	.	K	.	.	.	.	.	.	N	R	M	R	N	.	.	.	P	.	.	.	.	.	.
RSVB/Leipzig/2023/19	.	.	.	.	.	.	.	.	R	.	.	R	M	R	.	.	.	.	.	.	.	.	.	.	.
RSVB/Leipzig/2023/20	.	.	.	.	.	.	.	.	R	.	.	R	M	R	.	.	.	.	.	.	.	.	.	.	.
RSVB/Leipzig/2023/21	.	.	.	.	.	.	.	.	R	.	.	R	M	R	.	.	.	.	.	.	.	.	.	.	.
RSVB/Leipzig/2023/24	.	.	.	.	.	.	.	.	R	.	.	R	M	R	.	.	.	.	.	.	.	.	.	.	.
RSVB/Leipzig/2023/25	.	I	.	.	.	.	.	.	.	.	N	R	M	R	N	.	.	.	P	.	.	.	K	.	.

Amino acids (AA) are numbered with regard to the F0 protein. Additionally, the F1 and F2 subunits are indicated. The consensus amino acid residues to which the isolates were compared were derived from the GB5.0.5a consensus sequence. The AA changes were mapped to the following regions or antigenic sites: SP, signal peptide; p27, 27 amino acid fragments; A, heptad repeat A; B, heptad repeat B; CT, cytoplasmatic tail; Ø, site Ø; I, II, V, sites I, II, V. Strains that were identical at the amino acid level are depicted as a pool and contain isolates of the following seasons: [a] two strains for 2017/2018; [b] seven for 2017/2018, five for 2018/2019, ten for 2019/2020, three for 2021/2022; [c] nine for 2021/2022, 12 for 2022/2023.

4. Discussion

Following the introduction of NPIs [20,40] and changes in social practices, such as mask wearing in public and increased work from home [39], introduced worldwide and in Germany [41–44] in the wake of the SARS-CoV-2 pandemic in 2020, the seasonal circulation of other respiratory pathogens, such as RSV, was altered tremendously. Nationwide lockdowns with the subsequent closing of schools, daycare centers, and cultural events led to an exceptional reduction in RSV activity in the winter of 2022/2021. NPIs were eased starting in Germany in May 2021 [45] and tied to the incidence of SARS-CoV-2 infections. However, a more progressive easing of regulations for children and a higher and prolonged adhesion to NPIs in the adult population may represent key contributors to the disparity in the resurgence of RSV infections. While the pediatric population was greatly affected in the 2021/2022 season [4], the majority of the adult population was first affected in the 2022/2023 season.

The start and end points of the pre-pandemic seasons are in line with studies that, for continental Europe, typically define a season with a start in December and an ending in April of the following year [39–41]. For Germany, in particular, a season is defined based on the positivity rate (PR) exceeding 5% of RSV real-time RT-PCR detection assays in children between 0 and 4 years of age, as assessed by the national outpatient sentinel surveillance [42] of the Robert Koch Institute (RKI). This age subgroup was chosen because of its high disease burden and clinical importance in order to more accurately correspond to the endemic circulation. According to this definition, the pre-pandemic seasons spanned between weeks 50 to 51 (December) and weeks 11 to 13 (March). The seasonal peaks were reported between weeks four and six (January/February). For the sake of this study, the main detection interval was defined to be framed by each two subsequent weeks with at least two RSV cases per week in order to compensate for a lower catchment population compared to a sentinel surveillance and to additionally compensate for singular cases that would otherwise inflate the analyzed interval. For different study designs and locations, deviating parameters may be needed to generate a robust and comparable detection interval.

This study presents analyses of adult patients that indicate a slightly delayed main detection interval between 2017 and 2020 starting at weeks 4 and 5 (January/February) and ending at weeks 14 to 18 (April/May), which was also confirmed by the seasonal reports of the RKI [46]. The same applies to the peak detections, which were reported between weeks three and six (January/February) for the sentinel cohort and were observed between weeks nine and 14 (February to April) in the adult cohort studied. This distinctive circulation characteristic may be attributed to two main contributing factors. First, the epidemic wave of infection may originate from the infant population and subsequently reach older populations, similar to what was reported for influenza virus infections [47]. Secondly, a delay between the peaks of RSV infections in the general population and hospitalizations may be considered.

The regular RSV circulation pattern observed in the pre-pandemic seasons was remarkably altered with the start of NPIs due to the then-emerging SARS-CoV-2 pandemic. After an abrupt end of the RSV circulation after March 2020, only a few cases were noted for the projected time span of the 2020/2021 season, corresponding to an absent respiratory season [4,46]. After a gradual alleviation of the introduced NPIs, the re-emergence of RSV occurred as a premature season of 2021/2022 and was mainly driven by infections in the pediatric population. According to the definition that is used in Germany, the endemic circulation of RSV started in week 35 (August) and ended in week 50 (December) of 2021, peaking 19 weeks before the usual average (week eight) in week 41 (October). The main detection interval of the studied adult population was observed between weeks 37 (September) and 46 (November) with a peak in week 40 (October), devoid of the temporal delay described above. Furthermore, RSV circulation preceded the lowest air temperatures of the winter of 2021/2022, underlining the unusual properties of this respiratory season.

When compared with the pre-pandemic seasons, a premature start and end were also noted for the 2022/2023 season; however, a shift towards a regular cycle was observed. According to the RKI, in Germany, the endemic circulation occurred between week 41 in 2022 (October) and week 3 in 2023 (January), peaking 13 weeks before the usual average in week 47 (November) [46]. A re-establishment of the delay in RSV detections in the adult cohort was confirmed with a main detection interval between week 47 (November) and week 13 (March). The peak of RSV infections was observed in week 1 (January), eight weeks before the earliest observed peak of one of the pre-pandemic seasons, namely, seasons 2018/2019 and 2019/2020.

The considerably increased amount of RSV cases and altered circulation may highlight the underestimated importance of pre-existing or waning immunity, even in the adult population, as the re-emergence of RSV in 2021 apparently was not caused by major antigenic shifts circumventing a pre-existing immune response [4]. More likely, it seems that an immunologic gap caused by the lack of RSV circulation during the period of intensified contact restrictions was created within the population [48]. Furthermore, the importance of waning immunity and its impact on circulation patterns was previously described for RSV [49] and other respiratory viruses, like enteroviruses [50,51] and influenza B virus [52]. In addition to the apparent differences in the observed amount of RSV cases after the year 2020, age was a major epidemiologic characteristic that differed within the study period. It has to be noted, that the hospitalization rate due to RSV was previously reported to be less variable than e.g. rates observed for influenza over successive seasons. Additionally, the hospitalization burden showed a peak for adults between the ages 65 and 74 [5]. However, while the median age of the study population was 66.5 years in the pre-pandemic seasons, the age group of 25–44-year-old adults was over-represented in the 2021/2021 season, resulting in a significantly lower median age of 49.5 [4], which is in line with observed perturbations in the age composition of adults reported in other studies [53–55]. In the 2022/2023 season, the median age of the study population converted back to the pre-pandemic average (65 years). Therefore, excluding 2021/2022, between 65 and 80% of the adults of a respective season were 60 years of age or older and belonged to the target population of the newly introduced RSV vaccines (see below) [23,56]. The reasons for this remarkable disparity in the epidemiological characteristics are likely to be multifactorial and also have implications for the observed differences in the clinical parameters between the pre-pandemic seasons and the 2022/2023 season.

Overall, the prevalence of comorbidities was very high in the studied adult cohort, underlining observations of increased susceptibility to severe disease outcomes subsequent to RSV infections and the hospitalization of high-risk populations [3,5,10,57]. Comorbidities that for RSV have previously been reported to be associated with an increased incidence rate among hospitalized patients or with severe disease progression are, i.e., COPD, asthma, coronary artery disease, congestive heart failure [58–61], hematologic malignancies and previous stem cell transplantations [6,7], and lung transplantations [8]. Reported incidences of comorbidities were as high as 87% of the study population [62], and mortality after an RSV LRTI was reported to be up to 50% for specific subgroups, i.e., patients after hematopoietic cell transplantation with severe immunodeficiency [7].

As the number of comorbidities may vary substantially between different study sites, even within the same study [32], an observed baseline level and composition of comorbidities may be highly specific for the respective region and patient collective that is treated in or connected to each study site, and thus a comparison with other studies at the same site seems warranted. A comparable high number of comorbidities was also observed for infections with other respiratory pathogens at the same hospital, namely, rhinovirus [34], parainfluenza virus type 3 [63], and influenza B virus [52]. The particularly high number of ICU admissions (roughly 25%) of the RSV cases studied was comparable to what was observed for influenza B virus (20–25.8%) [52] and highlights the susceptibility of this patient cohort of developing a severe RSV infection. A comparison between the clinical characteristics of the pre-pandemic seasons and the 2022/2023 season may allow further conclusions about the patient collective at each point in time. Additionally, it may indicate behavioral changes evoked through the SARS-CoV-2 pandemic, i.e., a more liberal utilization of respiratory pathogen tests, including multiplex tests, because of an increased need for pathogen identification and higher awareness of severe viral infections.

After the re-emergence of RSV in 2021/2022, only a small number of adult patients tested positive for RSV at the study site. This may be attributed to an overall lower amount of RSV infections but also to the adherence of severely ill patients to the hospital, as indicated by the high number of immunosuppressed individuals [4]. Unfortunately, the low number of RSV cases in 2021/2022 was also the main obstacle to a robust comparison of clinical data with the two other periods of the current study. In the 2022/2023 season, the study population resembled the pre-pandemic baseline. However, subtle differences were noted. Firstly, the lower prevalence of RSV patients with a history of lung transplantation and immunosuppression may be the result of more stringent adherence to NPIs in specific and at-risk subgroups of adult patients [64]. The same may apply to patients with different stages of chronic kidney injury. Another reason may be indicated by the higher incidences of clinical features like LRTI, pneumonia, exacerbation of OLD, and administration of bronchodilators. As the incidence waves for RSV, influenza, and SARS-CoV-2 infections vastly overlapped in the 2022/2023 season [65], high pressure on the healthcare system was exerted, which may have resulted in a competition for resources and a selection for a higher disease severity in the hospitalized population. However, a waning immunity may also be considered to have an influence on the observed differences in clinical disease severity (see above).

The year 2023 may mark the beginning of a new era of dealing with RSV infections as new highly potent tools become implemented [66] in response to unmet needs for the prophylaxis of otherwise healthy newborns and adults. The vaccine efficacies for adults ≥ 60 years of age in the first RSV season were thereby reported to be up to 85.7% and 94.1%, depending on specific subgroups, for the bivalent [23] and monovalent [56] compositions, respectively. The protection rates for newborns for medically attended severe LRTIs following a maternal immunization were reported to be 81.8% and 69.4%, 90 and 180 days after birth, respectively [22]. For nirsevimab, the efficacy to prevent medically attended RSV-associated LRTI is reported to be 74.5% [22]. Additionally, the SARS-CoV-2 pandemic marked an evolutionary bottleneck, accelerating the notion of a very narrow spectrum of circulating genotypes and lineages [4,16,21], namely, the very closely related lineages GA2.3.5 and GA2.3.6a for RSV-A and GB5.0.5a for RSV-B. Thus, the timing for the introduction of new vaccines and long-acting monoclonal antibodies appears to be optimal.

At the same time, the implementation of new immunization strategies mainly targeting one antigenic site in a broad age range of the population may result in a selection of RSV strains with escape mutations in the fusion protein. In in vitro studies, prolonged selective pressure due to co-cultivation resulted in the selection of escape mutants with decreased binding affinities for various monoclonal antibodies (mABs) [67–71], including palivizumab [72,73]. Furthermore, resistant variants were detected in up to 8.7% of palivizumab recipients [74,75]. Thereby, resistance may not only be conferred by mutations at the binding site but also through distant mutations, resulting in an altered protein conformation [76]. Although generally considered a very conserved protein, the RSV-B isolates showed a greater diversity and drift in the fusion protein than the RSV-A isolates, confirming previous observations of accented differences in antigenic sites of the pre-fusion F [77]. Especially AAs that previously have been ascribed to site Ø [25] were affected in most RSV-B strains, and AA changes accumulated over the study period. In contrast, site II was affected to a much lower extent, and mutations rather appeared in clusters. A continued antigenic drift may have future implications for the efficacy of mABs and vaccines, especially in seasons with a predominance of RSV-B. Thus, molecular and antigenic surveillance of vaccine and mAB non-responders is highly warranted in the future. Furthermore, alternative vaccine strategies [78] and mABs [79] targeting different antigenic sites are needed in order to broaden the immunologic repertoire of the population.

Several limitations of this study should be noted. Notably, the genotyping approach was not applied to all RSV cases and may be a source of selection bias, which might include the circulation of other less common genotypes, especially in the seasons before the SARS-CoV-2 pandemic. However, the authors are confident that the analyzed RSV isolates constitute an adequate representation of the local F gene diversity in the study period and that the protocol applied was usable for contemporary analysis of the F gene, as a narrow genotype spectrum was demonstrated before [4]. Due to the retrospective study design, only associations could be shown, without proof of causality. Patient selection favoring severe cases may have occurred due to the sampling at a tertiary care hospital, which is underlined by the high numbers of pneumonia cases and cases that needed an ICU admission. In addition, the age spectrum of the analyzed patients is likely to underrepresent patients who do not belong to the high-risk groups for severe disease courses and hospitalization. Detection of a further pathogen was considered a co-infection; however, for bacterial pathogens, colonization cannot be ruled out.

5. Conclusions

This study reports the epidemiology and associated clinical spectrum of adult RSV cases that were treated at a tertiary care university hospital in Germany between 2017 and 2023. The findings are consistent with other studies and indicate a profound impact of non-pharmaceutical interventions implemented in the wake of the SARS-CoV-2 pandemic. A close phylogenetic relatedness of circulating strains is evidenced by analysis of the F gene. However, an accumulation of amino acid changes was observed, especially for RSV-B strains, and also affected immunologically important antigenic sites. The ongoing implementation of a recently approved monoclonal antibody with an extended half-life [24,30] and vaccination strategies for pregnant women and older adults [22,23,56] in clinical treatment guidelines necessitate a genetic and antigenic analysis of the fusion protein going forward. Additional epidemiologic studies or population-based surveillance programs are warranted, as retained behavioral changes are likely to have an influence on the re-establishment of a regular circulation of respiratory pathogens.

Supplementary Materials: The following supporting information can be downloaded at https://www.mdpi.com/article/10.3390/v16060943/s1, Table S1: Primers used for RSV (F gene) sequencing, Table S2: Reagent composition for RSV-A sequencing (first PCR), Table S3: Reagent composition for RSV-B sequencing (first PCR), Table S4: Cycling conditions for RSV sequencing (first PCR), Table S5: Reagent composition for RSV-A sequencing (nested PCR), Table S6: Reagent composition for RSV-B sequencing (nested PCR), Table S7: Cycling conditions for RSV sequencing (nested PCR), Table S8: RSV-A reference sequences by Goya et al., Table S9: RSV-B reference sequences by Goya et al., Table S10: Amino acid residues of antigenic sites Ø–V, Table S11: Co-infecting pathogens, Table S12: Study population and clinical features of RSV cases in the 2021/2022 and 2022/2023 seasons, Figure S1: Molecular phylogenetic analysis of the RSV-A F gene by the maximum likelihood method, Figure S2: Molecular phylogenetic analysis of the RSV-B F gene by the maximum likelihood method.

Author Contributions: Conceptualization, M.H. and C.P.; data curation, M.H.; formal analysis, M.H.; investigation, M.H.; methodology, M.H., A.F., S.T., S.B., M.M., T.C.W. and V.M.; resources, C.P. and C.L.; validation, M.H.; visualization, M.H. and S.T.; writing—original draft, M.H.; writing—review and editing, M.M., A.F., S.T., S.B., T.C.W., V.M., C.L. and C.P. All authors have read and agreed to the published version of the manuscript.

Funding: This research received no external funding.

Institutional Review Board Statement: The study was conducted in accordance with the Declaration of Helsinki and the protocol was approved by the Ethics Committee of Leipzig University (12 July 2021, No. Az 168/22-ek, and 21 November 2023, No. Az 360/23-ek).

Informed Consent Statement: This non-interventional study included no additional procedures. Anonymized biological material was only obtained for standard viral diagnosis. Informed consent for the storage and further use of the samples was obtained from all patients.

Data Availability Statement: Identified sequences were submitted to GenBank (accession No. PP084746-PP084890 and PP354071-PP354073).

Acknowledgments: The authors acknowledge the support from Leipzig University for open-access publishing. A. Frille was supported by the postdoctoral fellowship "MetaRot program" (clinician scientist program) from the Federal Ministry of Education and Research (BMBF), Germany (FKZ 01EO1501, IFB Adiposity Diseases), a research grant from the "Mitteldeutsche Gesellschaft für Pneumologie (MDGP) e.V." (2018-MDGP-PA-002), a junior research grant from the Medical Faculty, Leipzig University (934100-012), a graduate fellowship from the "Novartis Foundation", and the "PETictCAC" project (ERAPerMed_324), which was funded with tax funds based on the budget passed by the Saxon State Parliament (Germany) under the frame of ERA PerMed (Horizon 2020 Research and Innovation Framework Programme of the European Commission Research Directorate-General, Grant Agreement No. 779282).

Conflicts of Interest: The authors declare no conflicts of interest.

References

1. Borchers, A.T.; Chang, C.; Gershwin, M.E.; Gershwin, L.J. Respiratory syncytial virus—A comprehensive review. *Clin. Rev. Allergy Immunol.* **2013**, *45*, 331–379. [CrossRef] [PubMed]
2. Krause, C.I. The ABCs of RSV. *Nurse Pract.* **2018**, *43*, 20–26. [CrossRef] [PubMed]
3. Hall, C.B.; Long, C.E.; Schnabel, K.C. Respiratory syncytial virus infections in previously healthy working adults. *Clin. Infect. Dis.* **2001**, *33*, 792–796. [CrossRef] [PubMed]
4. Hönemann, M.; Thiem, S.; Bergs, S.; Berthold, T.; Propach, C.; Siekmeyer, M.; Frille, A.; Wallborn, T.; Maier, M.; Pietsch, C. In-Depth Analysis of the Re-Emergence of Respiratory Syncytial Virus at a Tertiary Care Hospital in Germany in the Summer of 2021 after the Alleviation of Non-Pharmaceutical Interventions Due to the SARS-CoV-2 Pandemic. *Viruses* **2023**, *15*, 877. [CrossRef] [PubMed]
5. Matias, G.; Taylor, R.; Haguinet, F.; Schuck-Paim, C.; Lustig, R.; Shinde, V. Estimates of hospitalization attributable to influenza and RSV in the US during 1997-2009, by age and risk status. *BMC Public Health* **2017**, *17*, 271. [CrossRef] [PubMed]
6. Khawaja, F.; Chemaly, R.F. Respiratory syncytial virus in hematopoietic cell transplant recipients and patients with hematologic malignancies. *Haematologica* **2019**, *104*, 1322–1331. [CrossRef]
7. Vakil, E.; Sheshadri, A.; Faiz, S.A.; Shah, D.P.; Zhu, Y.; Li, L.; Kmeid, J.; Azzi, J.; Balagani, A.; Bashoura, L.; et al. Risk factors for mortality after respiratory syncytial virus lower respiratory tract infection in adults with hematologic malignancies. *Transpl. Infect. Dis.* **2018**, *20*, e12994. [CrossRef]
8. Khalifah, A.P.; Hachem, R.R.; Chakinala, M.M.; Schechtman, K.B.; Patterson, G.A.; Schuster, D.P.; Mohanakumar, T.; Trulock, E.P.; Walter, M.J. Respiratory viral infections are a distinct risk for bronchiolitis obliterans syndrome and death. *Am. J. Respir. Crit. Care Med.* **2004**, *170*, 181–187. [CrossRef]
9. Falsey, A.R.; McElhaney, J.E.; Beran, J.; van Essen, G.A.; Duval, X.; Esen, M.; Galtier, F.; Gervais, P.; Hwang, S.-J.; Kremsner, P.; et al. Respiratory syncytial virus and other respiratory viral infections in older adults with moderate to severe influenza-like illness. *J. Infect. Dis.* **2014**, *209*, 1873–1881. [CrossRef]
10. Geevarghese, B.; Weinberg, A. Cell-mediated immune responses to respiratory syncytial virus infection: Magnitude, kinetics, and correlates with morbidity and age. *Hum. Vaccin. Immunother.* **2014**, *10*, 1047–1056. [CrossRef]
11. Griffiths, C.; Drews, S.J.; Marchant, D.J. Respiratory Syncytial Virus: Infection, Detection, and New Options for Prevention and Treatment. *Clin. Microbiol. Rev.* **2017**, *30*, 277–319. [CrossRef]
12. Hornung, F.; Rogal, J.; Loskill, P.; Löffler, B.; Deinhardt-Emmer, S. The Inflammatory Profile of Obesity and the Role on Pulmonary Bacterial and Viral Infections. *Int. J. Mol. Sci.* **2021**, *22*, 3456. [CrossRef]
13. Frydrych, L.M.; Bian, G.; O'Lone, D.E.; Ward, P.A.; Delano, M.J. Obesity and type 2 diabetes mellitus drive immune dysfunction, infection development, and sepsis mortality. *J. Leukoc. Biol.* **2018**, *104*, 525–534. [CrossRef] [PubMed]
14. Tabor, D.E.; Fernandes, F.; Langedijk, A.C.; Wilkins, D.; Lebbink, R.J.; Tovchigrechko, A.; Ruzin, A.; Kragten-Tabatabaie, L.; Jin, H.; Esser, M.T.; et al. Global Molecular Epidemiology of Respiratory Syncytial Virus from the 2017-2018 INFORM-RSV Study. *J. Clin. Microbiol.* **2020**, *59*, e01828-20. [CrossRef] [PubMed]
15. Trovão, N.S.; Khuri-Bulos, N.; Tan, Y.; Puri, V.; Shilts, M.H.; Halpin, R.A.; Fedorova, N.B.; Nelson, M.I.; Halasa, N.; Das, S.R. Molecular characterization of respiratory syncytial viruses circulating in a paediatric cohort in Amman, Jordan. *Microb. Genom.* **2021**, *7*, 000292. [CrossRef] [PubMed]
16. Goya, S.; Galiano, M.; Nauwelaers, I.; Trento, A.; Openshaw, P.J.; Mistchenko, A.S.; Zambon, M.; Viegas, M. Toward unified molecular surveillance of RSV: A proposal for genotype definition. *Influenza Other Respir. Viruses* **2020**, *14*, 274–285. [CrossRef] [PubMed]
17. Muñoz-Escalante, J.C.; Comas-García, A.; Bernal-Silva, S.; Noyola, D.E. Respiratory syncytial virus B sequence analysis reveals a novel early genotype. *Sci. Rep.* **2021**, *11*, 3452. [CrossRef] [PubMed]
18. Muñoz-Escalante, J.C.; Comas-García, A.; Bernal-Silva, S.; Robles-Espinoza, C.D.; Gómez-Leal, G.; Noyola, D.E. Respiratory syncytial virus A genotype classification based on systematic intergenotypic and intragenotypic sequence analysis. *Sci. Rep.* **2019**, *9*, 20097. [CrossRef]
19. Ma, Y.; Jiang, H.; Wan, Z.; Li, S.; Li, Y.; Wang, W.; Jin, X.; Li, Y.; Zhang, C. Evolutionary dynamics of group A and B respiratory syncytial virus in China, 2009-2018. *Arch. Virol.* **2021**, *166*, 2407–2418. [CrossRef]
20. Altman, G.; Ahuja, J.; Monrad, J.T.; Dhaliwal, G.; Rogers-Smith, C.; Leech, G.; Snodin, B.; Sandbrink, J.B.; Finnveden, L.; Norman, A.J.; et al. A dataset of non-pharmaceutical interventions on SARS-CoV-2 in Europe. *Sci. Data* **2022**, *9*, 145. [CrossRef]
21. Eden, J.-S.; Sikazwe, C.; Xie, R.; Deng, Y.-M.; Sullivan, S.G.; Michie, A.; Levy, A.; Cutmore, E.; Blyth, C.C.; Britton, P.N.; et al. Off-season RSV epidemics in Australia after easing of COVID-19 restrictions. *Nat. Commun.* **2022**, *13*, 2884. [CrossRef] [PubMed]
22. Kampmann, B.; Madhi, S.A.; Munjal, I.; Simões, E.A.F.; Pahud, B.A.; Llapur, C.; Baker, J.; Pérez Marc, G.; Radley, D.; Shittu, E.; et al. Bivalent Prefusion F Vaccine in Pregnancy to Prevent RSV Illness in Infants. *N. Engl. J. Med.* **2023**, *388*, 1451–1464. [CrossRef] [PubMed]
23. Walsh, E.E.; Pérez Marc, G.; Zareba, A.M.; Falsey, A.R.; Jiang, Q.; Patton, M.; Polack, F.P.; Llapur, C.; Doreski, P.A.; Ilangovan, K.; et al. Efficacy and Safety of a Bivalent RSV Prefusion F Vaccine in Older Adults. *N. Engl. J. Med.* **2023**, *388*, 1465–1477. [CrossRef] [PubMed]

24. Hammitt, L.L.; Dagan, R.; Yuan, Y.; Baca Cots, M.; Bosheva, M.; Madhi, S.A.; Muller, W.J.; Zar, H.J.; Brooks, D.; Grenham, A.; et al. Nirsevimab for Prevention of RSV in Healthy Late-Preterm and Term Infants. *N. Engl. J. Med.* **2022**, *386*, 837–846. [CrossRef] [PubMed]
25. Mas, V.; Nair, H.; Campbell, H.; Melero, J.A.; Williams, T.C. Antigenic and sequence variability of the human respiratory syncytial virus F glycoprotein compared to related viruses in a comprehensive dataset. *Vaccine* **2018**, *36*, 6660–6673. [CrossRef]
26. McLellan, J.S.; Chen, M.; Leung, S.; Graepel, K.W.; Du, X.; Yang, Y.; Zhou, T.; Baxa, U.; Yasuda, E.; Beaumont, T.; et al. Structure of RSV fusion glycoprotein trimer bound to a prefusion-specific neutralizing antibody. *Science* **2013**, *340*, 1113–1117. [CrossRef]
27. McLellan, J.S.; Yang, Y.; Graham, B.S.; Kwong, P.D. Structure of respiratory syncytial virus fusion glycoprotein in the postfusion conformation reveals preservation of neutralizing epitopes. *J. Virol.* **2011**, *85*, 7788–7796. [CrossRef]
28. Gilman, M.S.A.; Castellanos, C.A.; Chen, M.; Ngwuta, J.O.; Goodwin, E.; Moin, S.M.; Mas, V.; Melero, J.A.; Wright, P.F.; Graham, B.S.; et al. Rapid profiling of RSV antibody repertoires from the memory B cells of naturally infected adult donors. *Sci. Immunol.* **2016**, *1*, eaaj1879. [CrossRef] [PubMed]
29. Johnson, S.; Oliver, C.; Prince, G.A.; Hemming, V.G.; Pfarr, D.S.; Wang, S.C.; Dormitzer, M.; O'Grady, J.; Koenig, S.; Tamura, J.K.; et al. Development of a humanized monoclonal antibody (MEDI-493) with potent in vitro and in vivo activity against respiratory syncytial virus. *J. Infect. Dis.* **1997**, *176*, 1215–1224. [CrossRef]
30. Zhu, Q.; McLellan, J.S.; Kallewaard, N.L.; Ulbrandt, N.D.; Palaszynski, S.; Zhang, J.; Moldt, B.; Khan, A.; Svabek, C.; McAuliffe, J.M.; et al. A highly potent extended half-life antibody as a potential RSV vaccine surrogate for all infants. *Sci. Transl. Med.* **2017**, *9*, eaaj1928. [CrossRef]
31. WHO. International Statistical Classification of Diseases and Related Health Problems (ICD): Version 10. Available online: https://icd.who.int/browse10/2019/en (accessed on 31 March 2024).
32. Charlson, M.E.; Pompei, P.; Ales, K.L.; MacKenzie, C.R. A new method of classifying prognostic comorbidity in longitudinal studies: Development and validation. *J. Chronic Dis.* **1987**, *40*, 373–383. [CrossRef] [PubMed]
33. Davison, A.J.; Dolan, A.; Akter, P.; Addison, C.; Dargan, D.J.; Alcendor, D.J.; McGeoch, D.J.; Hayward, G.S. The human cytomegalovirus genome revisited: Comparison with the chimpanzee cytomegalovirus genome. *J. Gen. Virol.* **2003**, *84*, 17–28. [CrossRef] [PubMed]
34. Golke, P.; Hönemann, M.; Bergs, S.; Liebert, U.G. Human Rhinoviruses in Adult Patients in a Tertiary Care Hospital in Germany: Molecular Epidemiology and Clinical Significance. *Viruses* **2021**, *1*, 2027. [CrossRef] [PubMed]
35. Espy, M.J.; Ross, T.K.; Teo, R.; Svien, K.A.; Wold, A.D.; Uhl, J.R.; Smith, T.F. Evaluation of LightCycler PCR for implementation of laboratory diagnosis of herpes simplex virus infections. *J. Clin. Microbiol.* **2000**, *38*, 3116–3118. [CrossRef] [PubMed]
36. Stöcher, M.; Leb, V.; Bozic, M.; Kessler, H.H.; Halwachs-Baumann, G.; Landt, O.; Stekel, H.; Berg, J. Parallel detection of five human herpes virus DNAs by a set of real-time polymerase chain reactions in a single run. *J. Clin. Virol.* **2003**, *26*, 85–93. [CrossRef] [PubMed]
37. Deutscher Wetterdienst. Mean Monthly Temperature Station 2932. Available online: https://www.dwd.de/EN/climate_environment/cdc/clis/clis_node.html (accessed on 15 December 2023).
38. Kumar, S.; Stecher, G.; Tamura, K. MEGA7: Molecular Evolutionary Genetics Analysis Version 7.0 for Bigger Datasets. *Mol. Biol. Evol.* **2016**, *33*, 1870–1874. [CrossRef]
39. Tamura, K.; Nei, M. Estimation of the number of nucleotide substitutions in the control region of mitochondrial DNA in humans and chimpanzees. *Mol. Biol. Evol.* **1993**, *10*, 512–526. [CrossRef] [PubMed]
40. Iezadi, S.; Gholipour, K.; Azami-Aghdash, S.; Ghiasi, A.; Rezapour, A.; Pourasghari, H.; Pashazadeh, F. Effectiveness of non-pharmaceutical public health interventions against COVID-19: A systematic review and meta-analysis. *PLoS ONE* **2021**, *16*, e0260371. [CrossRef] [PubMed]
41. Bundesministerium der Justiz. Gesetzestext IfSG [German]. Available online: http://www.gesetze-im-internet.de/ifsg/index.html (accessed on 21 January 2023).
42. Bundesministerium für Gesundheit. COVID-19-SchG [German]. Available online: https://www.bundesgesundheitsministerium.de/service/gesetze-und-verordnungen/detail/gesetz-zur-staerkung-des-schutzes-der-bevoelkerung-und-insbesondere-vulnerabler-personengruppen-vor-covid-19.html (accessed on 21 January 2023).
43. Robert Koch Institut. Zusammenstellung IfSG [German]. Available online: https://www.rki.de/DE/Content/Infekt/IfSG/Gesetze/gesetze_node.html (accessed on 21 January 2023).
44. State Government of Saxony. Regulations Saxony [German]. Available online: https://www.coronavirus.sachsen.de/archiv-der-abgelaufenen-amtlichen-bekanntmachungen-7295.html (accessed on 21 January 2023).
45. Jørgensen, F.; Bor, A.; Rasmussen, M.S.; Lindholt, M.F.; Petersen, M.B. Pandemic fatigue fueled political discontent during the COVID-19 pandemic. *Proc. Natl. Acad. Sci. USA* **2022**, *119*, e2201266119. [CrossRef]
46. Robert Koch Institut. Weekly Reports of the AGI [German]. Available online: https://influenza.rki.de/Wochenberichte.aspx (accessed on 6 March 2023).
47. Greer, A.L.; Tuite, A.; Fisman, D.N. Age, influenza pandemics and disease dynamics. *Epidemiol. Infect.* **2010**, *138*, 1542–1549. [CrossRef]
48. Reicherz, F.; Xu, R.Y.; Abu-Raya, B.; Majdoubi, A.; Michalski, C.; Golding, L.; Stojic, A.; Vineta, M.; Granoski, M.; Cieslak, Z.; et al. Waning Immunity Against Respiratory Syncytial Virus During the Coronavirus Disease 2019 Pandemic. *J. Infect. Dis.* **2022**, *226*, 2064–2068. [CrossRef] [PubMed]

49. Stensballe, L.G.; Ravn, H.; Kristensen, K.; Meakins, T.; Aaby, P.; Simoes, E.A.F. Seasonal variation of maternally derived respiratory syncytial virus antibodies and association with infant hospitalizations for respiratory syncytial virus. *J. Pediatr.* **2009**, *154*, 296–298. [CrossRef] [PubMed]
50. Barclay, W.S.; al-Nakib, W.; Higgins, P.G.; Tyrrell, D.A. The time course of the humoral immune response to rhinovirus infection. *Epidemiol. Infect.* **1989**, *103*, 659–669. [CrossRef] [PubMed]
51. Pons-Salort, M.; Grassly, N.C. Serotype-specific immunity explains the incidence of diseases caused by human enteroviruses. *Science* **2018**, *361*, 800–803. [CrossRef] [PubMed]
52. Hönemann, M.; Martin, D.; Pietsch, C.; Maier, M.; Bergs, S.; Bieck, E.; Liebert, U.G. Influenza B virus infections in Western Saxony, Germany in three consecutive seasons between 2015 and 2018: Analysis of molecular and clinical features. *Vaccine* **2019**, *37*, 6550–6557. [CrossRef] [PubMed]
53. Falsey, A.R.; Cameron, A.; Branche, A.R.; Walsh, E.E. Perturbations in Respiratory Syncytial Virus Activity During the SARS-CoV-2 Pandemic. *J. Infect. Dis.* **2022**, *227*, 83–86. [CrossRef] [PubMed]
54. Panatto, D.; Domnich, A.; Lai, P.L.; Ogliastro, M.; Bruzzone, B.; Galli, C.; Stefanelli, F.; Pariani, E.; Orsi, A.; Icardi, G. Epidemiology and molecular characteristics of respiratory syncytial virus (RSV) among italian community-dwelling adults, 2021/22 season. *BMC Infect. Dis.* **2023**, *23*, 134. [CrossRef] [PubMed]
55. Buda, S.; Dürrwald, R.; Biere, B.; Reiche, J.; Buchholz, U.; Tolksdorf, K.; Schilling, J.; Goerlitz, L.; Streib, V.; Preuß, U.; et al. *ARE-Wochenbericht*, 2021.
56. Papi, A.; Ison, M.G.; Langley, J.M.; Lee, D.-G.; Leroux-Roels, I.; Martinon-Torres, F.; Schwarz, T.F.; van Zyl-Smit, R.N.; Campora, L.; Dezutter, N.; et al. Respiratory Syncytial Virus Prefusion F Protein Vaccine in Older Adults. *N. Engl. J. Med.* **2023**, *388*, 595–608. [CrossRef] [PubMed]
57. Nam, H.H.; Ison, M.G. Respiratory syncytial virus infection in adults. *BMJ* **2019**, *366*, l5021. [CrossRef]
58. Falsey, A.R.; Formica, M.A.; Hennessey, P.A.; Criddle, M.M.; Sullender, W.M.; Walsh, E.E. Detection of respiratory syncytial virus in adults with chronic obstructive pulmonary disease. *Am. J. Respir. Crit. Care Med.* **2006**, *173*, 639–643. [CrossRef]
59. Kurai, D.; Saraya, T.; Ishii, H.; Takizawa, H. Virus-induced exacerbations in asthma and COPD. *Front. Microbiol.* **2013**, *4*, 293. [CrossRef] [PubMed]
60. Falsey, A.R.; Walsh, E.E.; Esser, M.T.; Shoemaker, K.; Yu, L.; Griffin, M.P. Respiratory syncytial virus–associated illness in adults with advanced chronic obstructive pulmonary disease and/or congestive heart failure. *J. Med. Virol.* **2018**, *91*, 65–71. [CrossRef] [PubMed]
61. Branche, A.R.; Saiman, L.; Walsh, E.E.; Falsey, A.R.; Sieling, W.D.; Greendyke, W.; Peterson, D.R.; Vargas, C.Y.; Phillips, M.; Finelli, L. Incidence of Respiratory Syncytial Virus Infection Among Hospitalized Adults, 2017–2020. *Clin. Infect. Dis.* **2022**, *74*, 1004–1011. [CrossRef] [PubMed]
62. Lee, N.; Lui, G.C.Y.; Wong, K.T.; Li, T.C.M.; Tse, E.C.M.; Chan, J.Y.C.; Yu, J.; Wong, S.S.M.; Choi, K.W.; Wong, R.Y.K.; et al. High morbidity and mortality in adults hospitalized for respiratory syncytial virus infections. *Clin. Infect. Dis.* **2013**, *57*, 1069–1077. [CrossRef] [PubMed]
63. Martin, D.; Hönemann, M.; Liebert, U.G. Dynamics of nosocomial parainfluenza virus type 3 and influenza virus infections at a large German University Hospital between 2012 and 2019. *Diagn. Microbiol. Infect. Dis.* **2021**, *99*, 115244. [CrossRef]
64. Levitt, E.E.; Gohari, M.R.; Syan, S.K.; Belisario, K.; Gillard, J.; DeJesus, J.; Levitt, A.; MacKillop, J. Public health guideline compliance and perceived government effectiveness during the COVID-19 pandemic in Canada: Findings from a longitudinal cohort study. *Lancet Reg. Health Am.* **2022**, *9*, 100185. [CrossRef]
65. ECDC/WHO. Influenza Virus Characterization: Summary Report, Europe, February 2023. Copenhagen and Stockholm: WHO Regional Office for Europe and European Centre for Disease Prevention and Control; 2023 Licence: CC BY 3.0 IGO. Available online: https://www.ecdc.europa.eu/sites/default/files/documents/Influenza-characterisation-February-2023.pdf (accessed on 1 April 2024).
66. Karron, R.A. RSV Illness in the Young and the Old—The Beginning of the End? *N. Engl. J. Med.* **2023**, *388*, 1522–1524. [CrossRef]
67. Arbiza, J.; Taylor, G.; López, J.A.; Furze, J.; Wyld, S.; Whyte, P.; Stott, E.J.; Wertz, G.; Sullender, W.; Trudel, M. Characterization of two antigenic sites recognized by neutralizing monoclonal antibodies directed against the fusion glycoprotein of human respiratory syncytial virus. *J. Gen. Virol.* **1992**, *73 Pt 9*, 2225–2234. [CrossRef]
68. Rossey, I.; Gilman, M.S.A.; Kabeche, S.C.; Sedeyn, K.; Wrapp, D.; Kanekiyo, M.; Chen, M.; Mas, V.; Spitaels, J.; Melero, J.A.; et al. Potent single-domain antibodies that arrest respiratory syncytial virus fusion protein in its prefusion state. *Nat. Commun.* **2017**, *8*, 14158. [CrossRef]
69. Wu, S.-J.; Schmidt, A.; Beil, E.J.; Day, N.D.; Branigan, P.J.; Liu, C.; Gutshall, L.L.; Palomo, C.; Furze, J.; Taylor, G.; et al. Characterization of the epitope for anti-human respiratory syncytial virus F protein monoclonal antibody 101F using synthetic peptides and genetic approaches. *J. Gen. Virol.* **2007**, *88*, 2719–2723. [CrossRef]
70. Zhu, Q.; Lu, B.; McTamney, P.; Palaszynski, S.; Diallo, S.; Ren, K.; Ulbrandt, N.D.; Kallewaard, N.; Wang, W.; Fernandes, F.; et al. Prevalence and Significance of Substitutions in the Fusion Protein of Respiratory Syncytial Virus Resulting in Neutralization Escape from Antibody MEDI8897. *J. Infect. Dis.* **2018**, *218*, 572–580. [CrossRef] [PubMed]
71. López, J.A.; Bustos, R.; Orvell, C.; Berois, M.; Arbiza, J.; García-Barreno, B.; Melero, J.A. Antigenic structure of human respiratory syncytial virus fusion glycoprotein. *J. Virol.* **1998**, *72*, 6922–6928. [CrossRef] [PubMed]

72. Zhao, X.; Chen, F.-P.; Sullender, W.M. Respiratory syncytial virus escape mutant derived in vitro resists palivizumab prophylaxis in cotton rats. *Virology* **2004**, *318*, 608–612. [CrossRef] [PubMed]
73. Zhao, X.; Chen, F.-P.; Megaw, A.G.; Sullender, W.M. Variable resistance to palivizumab in cotton rats by respiratory syncytial virus mutants. *J. Infect. Dis.* **2004**, *190*, 1941–1946. [CrossRef] [PubMed]
74. Zhu, Q.; McAuliffe, J.M.; Patel, N.K.; Palmer-Hill, F.J.; Yang, C.; Liang, B.; Su, L.; Zhu, W.; Wachter, L.; Wilson, S.; et al. Analysis of Respiratory Syncytial Virus Preclinical and Clinical Variants Resistant to Neutralization by Monoclonal Antibodies Palivizumab and/or Motavizumab. *J. Infect. Dis.* **2011**, *203*, 674–682. [CrossRef] [PubMed]
75. Papenburg, J.; Carbonneau, J.; Hamelin, M.-È.; Isabel, S.; Bouhy, X.; Ohoumanne, N.; Déry, P.; Paes, B.A.; Corbeil, J.; Bergeron, M.G.; et al. Molecular Evolution of Respiratory Syncytial Virus Fusion Gene, Canada, 2006–2010. *Emerg. Infect. Dis.* **2012**, *18*, 120–124. [CrossRef] [PubMed]
76. Oraby, A.K.; Stojic, A.; Elawar, F.; Bilawchuk, L.; McClelland, R.; Erwin, K.; Granoski, M.; Griffiths, C.; Arutyunova, E.; Lemieux, M.J.; et al. A Single Amino Acid Mutation Alters the Neutralization Epitopes in the Respiratory Syncytial Virus Fusion Glycoprotein. **2024**, *Resarch Square preprint*.
77. Hause, A.M.; Henke, D.M.; Avadhanula, V.; Shaw, C.A.; Tapia, L.I.; Piedra, P.A. Sequence variability of the respiratory syncytial virus (RSV) fusion gene among contemporary and historical genotypes of RSV/A and RSV/B. *PLoS ONE* **2017**, *12*, e0175792. [CrossRef] [PubMed]
78. Schaerlaekens, S.; Jacobs, L.; Stobbelaar, K.; Cos, P.; Delputte, P. All Eyes on the Prefusion-Stabilized F Construct, but Are We Missing the Potential of Alternative Targets for Respiratory Syncytial Virus Vaccine Design? *Vaccines* **2024**, *12*, 97. [CrossRef]
79. Phuah, J.Y.; Maas, B.M.; Tang, A.; Zhang, Y.; Caro, L.; Railkar, R.A.; Swanson, M.D.; Cao, Y.; Li, H.; Roadcap, B.; et al. Quantification of clesrovimab, an investigational, half-life extended, anti-respiratory syncytial virus protein F human monoclonal antibody in the nasal epithelial lining fluid of healthy adults. *Biomed. Pharmacother.* **2023**, *169*, 115851. [CrossRef]

viruses

MDPI

Article

Genomic Evolution and Surveillance of Respiratory Syncytial Virus during the 2023–2024 Season

Madeline Yunker [1], Amary Fall [1], Julie M. Norton [1], Omar Abdullah [1], David A. Villafuerte [1], Andrew Pekosz [2,3], Eili Klein [2,4] and Heba H. Mostafa [1,*]

1 Johns Hopkins School of Medicine, Department of Pathology, Division of Medical Microbiology, Meyer B-121F, 600 N. Wolfe St., Baltimore, MD 21287, USA; myunker1@jhmi.edu (M.Y.); afall2@jhmi.edu (A.F.); jnorto19@jhmi.edu (J.M.N.); oabdull1@jhmi.edu (O.A.); dvillaf2@jh.edu (D.A.V.)
2 Department of Emergency Medicine, Johns Hopkins School of Medicine, Baltimore, MD 21287, USA; apekosz@jhsph.edu (A.P.); eklein@jhu.edu (E.K.)
3 W. Harry Feinstone Department of Molecular Microbiology and Immunology, The Johns Hopkins Bloomberg School of Public Health, Baltimore, MD 21287, USA
4 Center for Disease Dynamics, Economics, and Policy, Washington, DC 20005, USA
* Correspondence: hmostaf2@jhmi.edu; Tel.: +1-410-955-5077

Abstract: Respiratory syncytial virus (RSV) is a significant cause of morbidity, particularly in infants. This study describes RSV genomic diversity and disease outcomes during the 2023–2024 season in the Johns Hopkins Hospital System (JHHS). Between August and December 2023, 406 patient samples were sequenced, showing that RSV-B GB5.0.5a was the dominant genotype detected. RSV-A genotype GA2.3.5 was detected less frequently. Metadata analysis of patient data revealed that, although RSV-B was more commonly detected, patients with RSV-A infections were more frequently hospitalized. Analysis of both the G- and F-genes revealed multiple amino acid substitutions in both RSV-A and RSV-B, with some positions within the F-protein that could be associated with evasion of antibody responses. Phylogenetic analysis revealed the genetic diversity of circulating GB5.0.5a and GA2.3.5 genotypes. This study serves as an important baseline for genomic surveillance of RSV within the JHHS and will assist in characterizing the impact of the newly approved RSV vaccines on RSV genomic evolution and the emergence of escape mutations.

Keywords: respiratory syncytial virus; RSV; surveillance; genomic sequencing

Citation: Yunker, M.; Fall, A.; Norton, J.M.; Abdullah, O.; Villafuerte, D.A.; Pekosz, A.; Klein, E.; Mostafa, H.H. Genomic Evolution and Surveillance of Respiratory Syncytial Virus during the 2023–2024 Season. *Viruses* **2024**, *16*, 1122. https://doi.org/10.3390/v16071122

Academic Editors: Emanuele Castagno, Irene Raffaldi and Mark R. Krystal

Received: 29 April 2024
Revised: 5 July 2024
Accepted: 9 July 2024
Published: 12 July 2024

1. Introduction

Respiratory syncytial virus (RSV) is a significant cause of lower respiratory tract infections, causing bronchiolitis and pneumonia in infants and young children globally [1,2]. By the age of one, approximately 60–70% of children are infected with RSV, with 2–3% requiring hospitalization [3]. RSV seasonality is largely dependent on geographical location and generally results in annual epidemics during the winter months in temperate climates and year-round transmission in (sub)tropical climates [1,4]. However, the timing and duration of epidemics can vary widely between and within countries [1,5]. RSV activity also changes at the local level, with season onset, peak, and decline varying between 0 and 5 weeks across the country, within the same state [5,6], and in tropical countries [5]. Although circulating genotypes can be similar between two neighboring cities, genetic variation between locations is notable [7]. This indicates that local surveillance systems can be valuable for understanding surges in disease, outbreaks, and seasonality [5]. The infrequency of RSV genomic and clinical data largely stems from immature surveillance systems, which mainly rely on pre-existing influenza surveillance [3,4]. Genomic RSV surveillance will facilitate an understanding of outbreaks, informed responses, and the ability to predict future epidemiological waves [4]. Additionally, baseline RSV genome characterization is expected to facilitate future investigations on the impact of the recently approved RSV vaccines on RSV genomic evolution and changes in associated disease severity. In this study, we characterized RSV genotypes of patients diagnosed at the Johns Hopkins Health System (JHHS) during the 2023/2024 season and provided a description of the associated clinical presentations and outcomes.

2. Materials and Methods

2.1. Study Population

Standard-of-care diagnostic RSV testing was conducted for both inpatients and outpatients across JHHS hospitals and outpatient practices. Testing for RSV A/B was performed with Cepheid (Sunnyvale, CA, USA) Xpert Xpress SARS-CoV-2/Flu/RSV test or Roche (Basel, Switzerland) ePlex Respiratory Pathogen Panels [8,9]. Overall, 52,343 tests were conducted for RSV between June 2023 and February 2024. Of these, 4.6% (2420/52,343) were positive for RSV. Out of the 2420 samples positive for RSV, 17.2% (417/2420) were randomly selected (convenience sample, based on availability, sufficient volume, and proper storage after the standard-of-care testing) for genome sequencing. Study samples were collected between August 2023 and December 2023. Clinical and demographic data were collected in bulk from the electronic health record systems as described previously [10]. RSV-related admissions were defined based on laboratory testing results: patients tested with the extended respiratory panel were assumed to be symptomatic, and patients who tested positive only for RSV prior to admission were assumed to be admitted for RSV-related illness. Patients who tested positive for other targets were not included as being admitted for RSV-related illness.

2.2. Nucleic Acid Extraction and Real-Time PCR

Viral nucleic acid was extracted from remnant patient samples using the Chemagic Viral RNA/DNA Kit following the manufacturer's instructions (Revvity (Waltham, MA, USA)), with 300 μL extracted for each sample and eluted into 60 μL volume. To distinguish between RSV-A and RSV-B and obtain cycle threshold (Ct) values, real-time PCR was performed using the Luna Universal Probe One-Step RT-qPCR Kit per the manufacturer's guidelines (New England BioLabs, Ipswich, MA, USA). Samples with Ct values 33 and above were excluded from additional analysis.

2.3. Whole-Genome Amplification and Sequencing

Whole-genome amplification was adapted from Wang et al. [11] by using 2 μL of first-round PCR products for the nested PCR. Library preparation for sequencing followed the manufacturer instructions of the DNA Native Barcoding Kit 96 v14 (SQK-NBD114.96) for the PromethION and the NEBNext ARTIC Library Prep kit (New England BioLabs/Oxford Nanopore Technologies (Oxford, UK)). The input amount of the PCR product used was 7 μL of the nested PCR product in addition to 4 μL of the first-round PCR product into the end-prep reaction. EDTA was added to the barcoded amplicons to stop the reaction. Samples were pooled and 5 μL/sample AMPure Beads were used for sample cleaning. Elution was added to 10 μL of NEBNext Quick Ligation Reaction Buffer, 5 μL of native adaptor, and 5 μL of Quick T4 DNA Ligase and incubated at room temperature for 20 min. Beads (30 μL) were added, and clean-up steps were repeated using 125 μL of SFB buffer. The library was eluted in 35 μL of elution buffer. The entire 35 μL library was used for sequencing.

2.4. Virus Genome Assembly and Phylogenetic Analysis

The resulting fastq files were analyzed using our in-house pipeline. The closest references were selected by blasting against the RSV reference genomes (NC_038235 and NC_001781). Draft genomes were generated by mini_assemble within pomoxis, using medaka consensus to further enhance the draft genome and establish a consensus sequence. Sequencing depth was evaluated with samtools. The alignment of sequences was performed using the built-in pipeline in NextClade. Clades and amino acid substitutions (AASs) were determined by the built-in pipeline in NextClade. Quality control parameters were used to remove sequences of mediocre and bad quality, with scores between 30 and 90+. Genomes were visualized using BioEdit and IGV to assess the presence of gaps and coverage. Sequences with gaps were removed from the analysis. The phylogenetic trees for RSV-A and RSV-B were generated using the maximum likelihood method using IQ-Tree version 2.2.6. The visualization was performed using FigTree version 1.4.4. Complete reference genomes used for the phylogenetic analysis can be found in Supplementary Table S1. The ModelFinder in IQ-TREE2 was used to select the best-fitted nucleotide substitution model. The robustness of the tree topology was tested with 1000 nonparametric bootstrap analyses. Bootstrap values > 75% were shown on branches of the consensus trees.

3. Results

3.1. RSV Prevalence at JHHS and the Study Cohort

The 2023/2024 respiratory viral season (≈12.5% at peak) exhibited a lower RSV positivity rate in the JHHS than the 2022/2023 (≈15% at peak) and 2021/2022 (≈20% at peak) seasons (Figure 1). The 2021/2022 RSV season at JHHS exhibited increasing testing positivity rates in early April 2021, peaking in September 2021 with a testing positivity rate close to 20% (Figure 1). Similarly, the 2022/2023 season exhibited increasing testing positivity rates in June 2022, peaking in October 2022 with a testing positivity rate of 12.8% (Figure 1). For the 2023/2024 season, increased RSV detection started around early September 2023 and peaked in November 2023 with a positivity rate of 10.9% (Figure 1). RSV activity peaked earlier in the season compared to influenza A and B viruses, consistent with the prior two seasons (Figure 1).

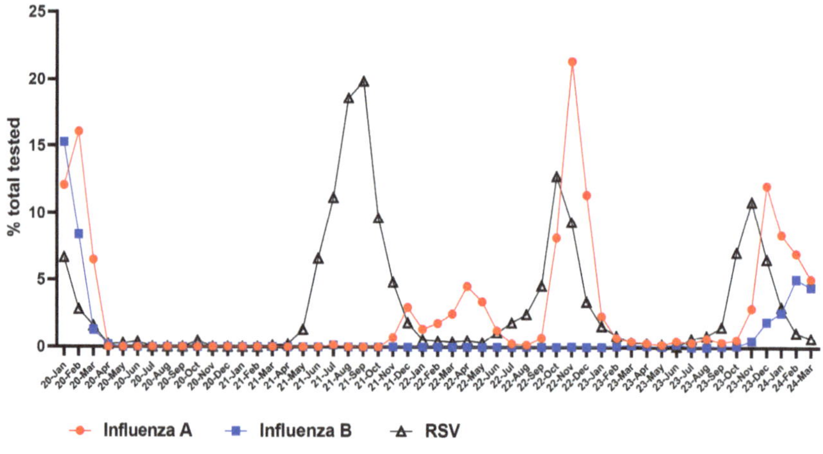

Figure 1. Percent positive tests (%) for influenza A, influenza B, and RSV between January 2020 and December 2023.

Metadata of the 2023/2024 cohort were collected for 406 unique patients. The cohort consisted of slightly more females (51.0%, [207/406]) than males (48.8%, [198/406]) (Table 1). The median age of the cohort was two years old. Children aged 1–5 years were the most represented (45.7%, [183/406]). Infants, 11 months and younger, comprised 29.6% (120/406) of the cohort. Adults 60 years and older (9.1%, [37/406]) and the other age groups were represented to a lesser extent (Table 1). Clinical signs and symptoms were noted from the charts of 89.4% (363/406) of patients at the time of presentation. Cough (32.8%, [119/363]), fever (28.9%, [105/363]), and breathing problems (27.0%, [98/363]), including wheezing and shortness of breath, were most commonly reported (Table 1). One or more comorbidities were noted for 40.1% (163/406) of patients, with cancer (23.9%, [97/406]), immunosuppression (20.7%, [84/406]), and hypertension (17.0%, [69/406]) being most common (Table 1). Admissions were reported for 31.3% (127/406) of the cohort; of those, 23.6% (30/127, and 7.4% of cohort, [30/406]) required ICU-level care (Table 1).

Supplemental oxygen was provided to 77.2% (98/127, and 24.1% of cohort [98/406]) of admitted patients (Table 1). The age groups 18–59 years and 60+ years had the highest rates of comorbidities and admissions, as detailed in Table S2. Metadata also revealed 13 (3.2%) patients to have co-infections with rhino/enteroviruses. These co-infections occurred in all age groups, with children 1–5 years of age and infants comprising 46.2% (6/13) and 23.1% (3/13), respectively. Other respiratory pathogens (influenza A/B, parainfluenza 1–4, seasonal coronaviruses, SARS-CoV-2, adenovirus, metapneumovirus, *Chlamydia pneumoniae*, and *Mycoplasma pneumoniae*) were included in the metadata pull but were not detected in our cohort.

Table 1. Description of 2023/2024 study cohort. Breathing problems include wheezing and shortness of breath. Seizures include febrile seizures. Percents for symptoms were calculated out of the total number of patients with symptoms reported in their chart. All others, unless indicated, were calculated out of the total cohort or samples with characterized genotypes.

Characteristic	**Number of Patients (% Cohort)**			
	Cohort	RSV B	RSV A	RSV A/B
Cohort Size	406	282 (69.5)	78 (19.2)	14 (3.4)
	Sex, N (%)			
Female	207 (51.0)	136 (48.2)	51 (65.4)	5 (35.7)
Male	198 (48.8)	146 (51.8)	26 (33.3)	9 (64.3)
Nonbinary	1 (0.2)	0	1 (1.3)	0
	Age Group, N (%)			
Infants (0–11 months)	120 (29.6)	82 (29.1)	27 (34.6)	6 (42.9)
1–5	183 (45.1)	133 (47.2)	28 (35.9)	7 (50.0)
6–17	28 (6.9)	19 (6.7)	4 (5.1)	0
18–59	38 (9.4)	21 (7.4)	12 (15.4)	1 (7.1)
≥60	37 (9.1)	27 (9.6)	7 (9.0)	0
	Clinical presentation, N (%)			
Symptomatic	363 (89.4)	255 (90.4)	65 (83.3)	12 (85.7)
Not Symptomatic	43 (10.6)	27 (9.6)	13 (16.7)	2 (14.3)
Fever	105 (28.9)	71 (27.8)	19 (29.2)	2 (16.7)
Cough	119 (32.8)	76 (29.8)	25 (38.5)	4 (33.3)
Breathing Problems	98 (27.0)	74 (29)	15 (23.1)	4 (33.3)
Congestion	21 (5.8)	12 (4.7)	6 (9.2)	1 (8.3)
Emesis	20 (5.5)	16 (6.3)	3 (4.6)	0
URI	20 (5.5)	15 (5.9)	4 (6.2)	1 (8.3)
Flu-like Symptoms	11 (3.0)	8 (3.1)	1 (1.5)	0
Chest Pain	7 (1.9)	5 (2.0)	2 (3.1)	0
Seizures	1 (0.3)	0	0	0
	Comorbidity, N (%)			
≥1 Underlying Medical Condition	163 (40.1)	108 (38.3)	32 (41.0)	9 (64.3)
Hypertension	69 (17.0)	42 (14.9)	15 (19.2)	3 (21.4)
Lung Disease	37 (9.1)	27 (9.6)	6 (7.7)	2 (14.3)
Kidney Disease	47 (11.6)	29 (10.3)	12 (15.4)	3 (21.4)

Table 1. *Cont.*

Characteristic	Number of Patients (% Cohort)			
Immunosuppression	84 (20.7)	56 (19.9)	15 (19.2)	4 (28.6)
Diabetes	26 (6.4)	14 (5.0)	10 (12.8)	0
Heart Failure	26 (6.4)	16 (5.7)	8 (10.3)	0
Atrial Fibrillation	19 (4.7)	14 (5.0)	3 (3.8)	1 (7.1)
Smoker	16 (3.9)	7 (2.5)	6 (7.7)	0
Cerebrovascular Disease	22 (5.4)	16 (5.7)	6 (7.7)	0
Cancer	97 (23.9)	61 (21.6)	25 (32.1)	4 (28.6)
Coronary Artery Disease	46 (11.3)	31 (11.0)	11 (14.1)	0
Pregnancy	5 (1.2)	1 (0.4)	2 (2.6)	0
	Outcome, N (%)			
Admitted	127 (31.3)	77 (27.3)	33 (42.3)	8 (57.1)
ICU	30 (7.4)	21 (7.4)	7 (9.0)	1 (7.1)
Supplemental Oxygen	98 (24.1)	63 (22.3)	23 (29.5)	6 (42.9)

Metadata were compared for 374 patients, whose samples were able to be typed for RSV-A, RSV-B, or RSV-A/B (Table 1). RSV-B (51.8%, [146/282]) and RSV-A/B (64.3%, [9/14]) comprised slightly more males than females, whereas the RSV-A cohort had slightly more females (65.4%, [51/78]). The median age for RSV-B, RSV-A, and RSV-A/B were similar. Of note, RSV-A had a higher percentage of adults 18-59 years of age (15.4%, [12/78]) compared to RSV-B (7.4%, [21/282]) and RSV-A/B (7.1%, [1/14]). Clinical signs were reported at the time of presentation in 90.4% (255/282) of RSV-B, 83.3% (65/78) of RSV-A, and 85.7% of RSV-A/B (12/14) and were similar between the three groups. RSV-A had higher admissions than RSV-B, with 42.3% (33/78) of RSV-A and 27.3% (77/282) of RSV-B patients being admitted (Table 1). A total of 27.3% (21/77) of admitted RSV-B patients and 21.2% (7/33) of admitted RSV-A patients required ICU-level care. Rhino/enterovirus co-infections were found in 3.2% (9/282) of RSV-B-positive samples and 3.8% (3/78) of RSV-A-positive samples. No co-infections in RSV-A/B samples were reported.

3.2. Genotype Analysis

Following real-time PCR for initial genotyping into RSV-A and RSV-B, 33 (7.9%) samples were excluded from whole-genome sequencing for having Ct values above 33. Of the remaining samples, 289 (69.3%) were identified as RSV-B, 81 (19.4%) were RSV-A, and 14 (3.4%) were RSV-A/B co-detections. A total of 255 (61.2%) sequences had complete G-genes and were assigned GA genotype classifications based on the Nextclade Pipeline. The complete F-gene was recovered from 266 (63.8%) sequences.

3.3. G-Gene Analysis

Of the 52 RSV-A samples with G-gene sequences, all (52/52) belonged to G-clade GA2.3.5. Several different clades were identified based on defining AASs in the G-gene (Table 2). A.D.1 comprised 38.5% (20/52), A.D.5.2 was identified in 28.8% (15/52), and A.D.3 in 19.2% (10/52) of samples (Table 2). A.D.2.1 (1.9%, [1/52]), A.D.3.1 (3.8%, [2/52]), and A.D.5.1 (7.7%, [4/52]) were represented to a lesser extent (Table 2).

Table 2. Distribution of G-clades and clades for RSV-A and RSV-B sequences obtained from the JHHS 2023 cohort.

	RSV-A	
	Number of Samples (% Cohort)	%
Total	52 (20.4)	
	G-Clade	
GA2.3.5	52	100
	Clades	
A.D.1	20	38.5
A.D.2.1	1	1.9
A.D.3	10	19.2
A.D.3.1	2	3.8
A.D.5.1	4	7.7
A.D.5.2	15	28.8
	RSV-B	
Total	203 (79.6)	
	G-Clade	
GB.5.0.5a	203	100
	Clades	
B.D.4.1	1	0.5
B.D.4.1.1	15	7.4
B.D.E.1	187	92.1

Of the RSV-B samples with G-gene sequences, all (203/203) belonged to the G-clade GB5.0.5a. Within this clade, three different clades were identified based on defining AASs in the G-gene (Table 2). B.D.E.1 represented 92.1% (187/203) of sequences and B.D.4.1.1 was found in 7.4% (15/203) of samples.

Twelve AASs were identified in the RSV-A G-gene that were found across all or most of the clades when compared to the reference sequence hRSV/A/England/397/2017 (Table 3). P71L, S243I, and I265L were found in 100% of RSV-A genomes. G224E (69.2%), D284G (96.2%), and Y304H (82.7%) were found across all clades but not all samples. H90Y (96.2%, not in A.D.2.1), L101F (88.5%, not in A.D.3.1), I134K (98.1%, not in A.D.2.1), and K262E (98.1%, not in A.D.2.1) were found in four of five clades. Two AASs were found in the majority of samples, T319I (69.2%) and T320A (73.1%) (Table 3). V131D (100%, [10/10]), I141T (70%, [7/10]), H266L (100%, [10/10]), and T136I (90%, [9/10]) were frequently identified within the A.D.3 clade.

Seven AASs were identified to be in most or all the RSV-B clades. A74V (100%), T131A (96.1%), I137T (97.0%), and I252T (90.6%) were found in all RSV-B clades when compared with the reference genome hRSV/B/Australia/VIC-RCH056/2019 (Table 3). Of note, I252T was only found in one (6.7%, [1/15]) B.D.4.1.1 sample. An additional AAS, P289L, was identified across all three clades, but at low percentages (2.5%, [5/203]). P289S (13.3%, [2/15]) was found in clade B.D.4.1.1. The B.D.4.1.1 and B.D.E.1 clades had two additional AASs that were found among the majority of samples. S100G was found in 98.5% (200/203) of the RSV-B samples, and all B.D.E.1 samples contained this substitution (Table 3). P221L was found in 96.6% (196/203) of the RSV-B samples. Three more substitutions were identified in low frequencies between both clades: P101S (B.D.4.1.1: 6.7%, [1/15], B.D.E.1: 1.6%, [3/187]), T141I (B.D.4.1.1: 6.7%, [1/15], B.D.E.1: 0.5%, [1/187]), and E239K (B.D.4.1.1: 13.3%, [2/15], B.D.E.1: 13.9%, [26/187]). A further five AASs were frequently found in B.D.4.1.1 samples: K85E (86.7%, [13/15]), T141A (80%, [12/15]), E224G (80%, [12/15]),

P229S (66.7%, [10/15]), and S243I (80%, [12/15]). The loci 85 (K85T) and 141 (T141I) also exhibited substitutions in two samples within the B.D.E.1 clade. Within the B.D.E.1 clade, four AASs were prominent: K256N (90.4%, [169/187]), I268T (98.9%, [185/187]), S275P (98.4%, [184/187]), and Y285H (78.1%, [146/187]). An additional 41 samples (21.9%) had a Y285L substitution.

Table 3. Amino acid substitutions identified within the G-region of RSV-A and RSV-B viruses in the JHHS 2023 cohort.

			RSV-A				
			Number of Samples with Substitution (%)				
Substitution	Total	A.D.1	A.D.2.1	A.D.3	A.D.3.1	A.D.5.1	A.D.5.2
P71L	52 (100)	20 (100)	1 (100)	10 (100)	2 (100)	4 (100)	15 (100)
H90Y	50 (96.2)	20 (100)	0	9 (90.0)	2 (100)	4 (100)	15 (100)
L101F	46 (88.5)	18 (90.0)	1 (100)	8 (80.0)	0	4 (100)	15 (100)
I134K	51 (98.1)	20 (100)	0	10 (100)	2 (100)	4 (100)	15 (100)
G224E	36 (69.2)	2 (15.0) G224V: 17 (85.0)	1 (100)	10 (100)	2 (100)	4 (100)	15 (100)
S243I	52 (100)	20 (100)	1 (100)	8 (80.0)	2 (100)	4 (100)	15 (100)
K262E	51 (98.1)	20 (100)	0	10 (100)	2 (100)	4 (100)	15 (100)
I265L	52 (100)	20 (100)	1 (100)	10 (100)	2 (100)	4 (100)	15 (100)
D284G	50 (96.2)	20 (100)	1 (100)	9 (90.0)	1 (50)	4 (100)	15 (100)
Y304H	43 (82.7)	20 (100)	1 (100)	2 (20.0)	1 (50)	4 (100)	15 (100)
T319I	36 (69.2)	17 (85.0)	T319S1 (100)	0	0	4 (100)	15 (100)
T320A	38 (73.1)	18 (90.0)	0	0	1 (50)	4 (100)	15 (100)
			RSV-B				
			Number of Samples with Substitution (%)				
Substitution	Total	B.D.4.1	B.D.4.1.1	B.D.E.1			
A74V	203 (100)	1 (100)	15 (100)	187 (100)			
T131A	195 (96.1)	1 (100)	13 (86.7)	181 (96.8)			
I137T	197 (97.0)	1 (100)	13 (86.7)	183 (97.9)			
I252T	184 (90.6)	1 (100)	1 (6.7)	182 (97.3)			
P289L	5 (2.5)	1 (100)	1 (6.7)	3 (1.6)			
S100G	200 (98.5)	0	13 (86.7)	187 (100)			
P221L	196 (96.6)	0	12 (80.0)	184 (98.4)			

3.4. F-Gene Analysis

The F-genes of 53 (65.4%) RSV-A sequences were examined for AASs. A T122A substitution was found in 43.4% of samples. This locus was not found to have any other substitutions with different amino acid identities within RSV-A samples. A V127I substitution was found in 24.5% of samples (Table 4). A V127A substitution was also found in 9.4% (5/53) of RSV-A samples. The F-genes of 213 (73.7%) RSV-B sequences were examined for AASs. An S211N substitution was detected in 96.2% (205/213) and an S389P substitution was detected in 93.0% (198/213) of RSV-B samples (Table 4). A further 4.2% (9/213) of samples had an S389L substitution.

Table 4. AASs identified within the F-region of RSV-A and RSV-B viruses in the JHHS 2023/2024 cohort.

	RSV-A	
Total	53	
Substitution	Number of Samples with Substitution	%
T122A	23	43.4
V127I	13	24.5
V127A	5	9.4
	RSV-B	
Total	213	
Substitution	Number of Samples with Substitution	%
S211N	205	96.2
S389P	198	93.0
S389L	9	4.2

AASs that have been reported to impact or are at positions with substitutions that have been reported to impact monoclonal antibody binding in both RSV-A and RSV-B F-proteins were detected. An S276N substitution was detected in 5.7% [3/53] of RSV-A samples (Table 5). Additionally, N201K (1.9%, [4/213]), R209Q (2.3%, [5/213]), and R209K (1.4%, [3/213]), which might impact palivizumab and nirsevimab activity, were detected [12].

Table 5. Amino acid substitutions in RSV-A and RSV-B F-region known to or * at positions known to impact monoclonal antibody binding.

Substitution	Number of Samples	%	Monoclonal Antibody Impacted
		RSV-A	
S276N	3	5.7 [3/53]	Palivizumab
		RSV-B	
N201K *	4	1.9 [4/213]	Nirsevimab
R209Q *	5	2.3 [5/213]	Nirsevimab
R209K *	3	1.4 [3/213]	Nirsevimab

3.5. Phylogeny

The Nextclade G-gene results were confirmed through a phylogenetic analysis for RSV-A (Figure 2) and RSV-B (Figure 3). For RSV-A, the clades described above (A.D.1, A.D.2.1, A.D.3, A.D.3.1, A.D.5.2, and A.D.5.2) clustered together with reference genomes from the GA2.3.5 clade. Each clade clustered with reference genomes from different geographical locations and isolation years (Table S1). The A.D.1 clade had two distinct branches, with reference genomes from the United States but from different years. The smaller cluster was closely related to a 2022 genome, whereas the larger cluster was from 2023. The two A.D.3 samples were found to cluster with the ten A.D.3.1 samples but were found to be within their own branch (Figure 2). Within the A.D.3 genotypes, there were three distinct branches with references from the United States (characterized in 2023 and 2022) and China (characterized in 2019) (Figure 2 and Table S1). The clade A.D.5.1 clustered with a 2022 reference genome from the United States. The largest clade, A.D.5.2, was close to 2022 reference genomes from the United States and two from Germany characterized in 2021 and 2022 (Figure 2 and Table S1). Genotype A.D.2.1 was most closely related to a 2022 reference genome from the United States (Table S1).

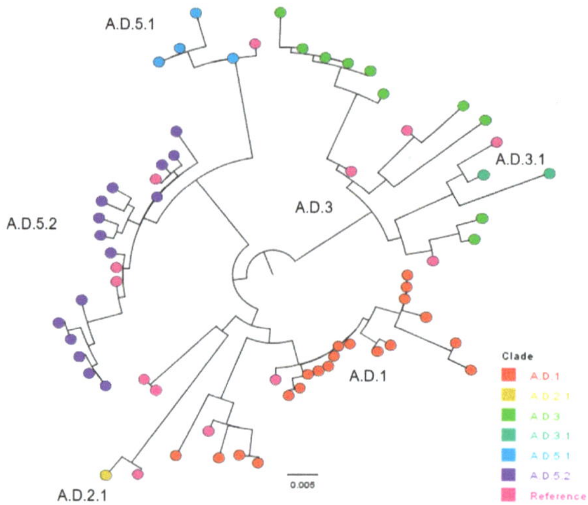

Figure 2. Phylogenetic tree for RSV-A G-gene. Reference genomes are listed in Table S1. All sequences belonged to the G-clade GA2.3.5.

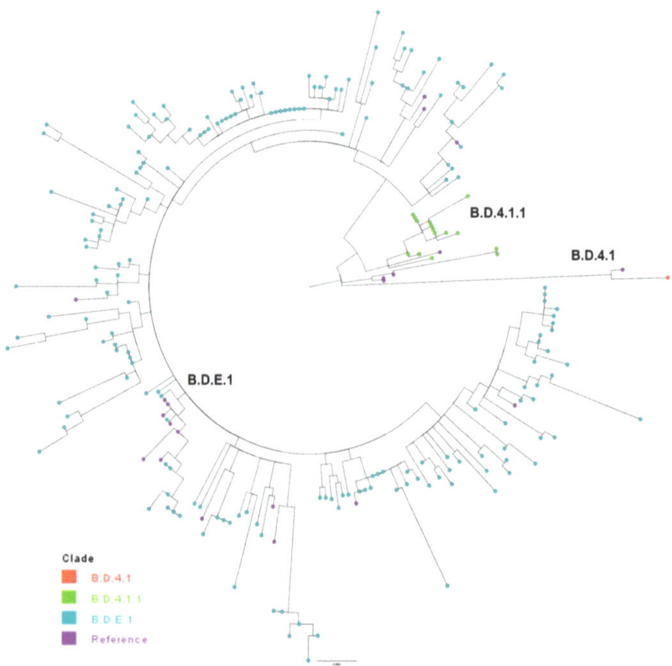

Figure 3. Phylogenetic tree for RSV-B G-gene. All clades belonged to the G-clade GB5.0.5a. Reference genomes are in purple.

For RSV-B, the clades B.D.4.1, B.D.4.1.1, and B.D.E.1 clustered distinctly, confirming the Nextclade analysis (Figure 3). All samples were associated with sequences belonging to the GB5.0.5a genotype. The clade B.D.4.1 clustered distinctly, along with a 2021 reference genome from France (Figure 3 and Table S1). The B.D.4.1.1 clade also clustered distinctly, with some evident diversity. These samples clustered with two geographically and temporally distinct reference genomes. Two samples were most closely related to a 2018 sample from the United States, while the majority clustered with a 2023 sample from Thailand. For

B.D.E.1, there were many distinct clusters indicating lots of diversity within this clade. The B.D.E.1 reference genomes were from three different countries in 2021–2024: United States, Thailand, and Germany (Figure 3 and Table S1).

4. Discussion

4.1. RSV-B Is the Dominant Genotype Characterized from the JHHS

In the years following the emergence of SARS-CoV-2, the seasonality of RSV was temporarily disrupted, as was the case for numerous respiratory pathogens [13]. RSV cases for the 2021/2022 season began in mid-April 2021 and peaked in September. This early beginning to the RSV season was also reported in Washington State [13]. The 2022/2023 season also exhibited an early start, with cases increases starting in July, peaking in October, and ending around February 2023. Washington State, Arizona, and Massachusetts also observed cases increasing and peaking earlier than normal [13–15]. Cases during the 2023/2024 season increased in September and peaked in mid-November. This is still earlier than what has been reported in the past, with the United States exhibiting peak RSV season in January/February [4]. Further retrospective studies of RSV circulation within the United States indicated that the season onset usually occurred in October/November, peaking in December/January, and decreasing around March/April [6,16]. However, it is important to note that RSV seasonality is prone to substantial differences at the national and subnational level [5].

The RSV testing positivity rate increased during the 2021/2022 season and decreased during the following two seasons. Our data contrast with those from Washington State, which reported a greater number of cases during the 2022/2023 season [13]. However, an increase in RSV cases in the 2021/2022 and 2022/2023 seasons was observed globally [17–20]. This increase in cases is thought to be due to reduced protective immunity following the COVID-19 pandemic, with population exposure to RSV having been impacted by non-pharmaceutical interventions (NPIs) [19,21]. Previous studies have shown that in the absence of circulating RSV, antibody titers waned significantly [20]. Studies have indicated reduced anti-RSV antibody levels in infants, women of childbearing age, and in human milk, indicating immune debt [17]. However, another study from Australia showed no differences in the population-level immunity to RSV between pre-pandemic seasons and the 2022 season [20]. It was also hypothesized that a new viral strain with increased fitness may have emerged following the pandemic [13,15]. However, numerous reports from Australia, Japan, Italy, Austria, Argentina, and Spain reported similar genotypes circulating pre- and post-pandemic, indicating that this surge in cases was likely not due to a new viral genotype [7,17,19–22]. In fact, there was a decrease in the genetic diversity of circulating RSV-A and RSV-B genotypes in Italy and Australia during and after the pandemic, indicating a potential genetic bottleneck introduced due to a sharp reduction in infections [17,20]. The increased number of cases following the pandemic might be attributable to more complex and unknown factors, such as changes in infrastructure, social attitudes, and health-seeking behaviors [20].

RSV-B genotype GB5.0.5a was predominantly (69.5%) detected in our cohort, followed by RSV-A genotype GA2.3.5 (19.2%), and a few co-detections of RSV-A/B (3.4%). In the United States, RSV-A was predominantly detected during the 2021/2022 and 2022/2023 seasons in Washington, Arizona, and Massachusetts [13–15]. In Austria and Bulgaria, RSV-B dominated the 2022/2023 season, while RSV-A drove the surge in 2021/2022 [19]. The RSV-B genotype GB5.0.5a comprised all RSV-B samples in Washington State during the previous seasons, whereas RSV-A genotype GA2.3.6b was dominant, with GA2.3.5 circulating to a lesser extent [13]. Similarly, in Arizona, RSV-B genotype GB5.0.5a comprised all RSV-B samples [14]. However, similar to what our data show, RSV-A GA2.3.5 was the only genotype detected. In Massachusetts, all RSV-B belonged to the GB5.0.5a genotype and all RSV-A belonged to the GA2.3.5 genotype [15]. The GB5.0.5a and GA2.3.5 genotypes have been circulating globally since 2017 and reported to be dominant following the SARS-CoV-2 pandemic [19,20,22]. However, some variability is observed among countries.

GA2.3.6b was found to be circulating in Argentina from 2019 to 2021 [21]. Our data indicate a shift in predominant circulating viruses from RSV-A to RSV-B in the 2023/2024 season, although the circulation of genotypes within RSV-A and RSV-B are similar to what was detected in previous seasons. RSV-A and RSV-B have been shown to co-circulate during epidemic seasons, with predominance altering over time [23]. The co-circulation of RSV subtypes is associated with co-detections of both RSV-A and RSV-B, comprising around 2% of cohorts in studies from the United States and Senegal [23,24]. The percentage of co-detections reported in our cohort was slightly higher; however, the clinical significance and outcomes associated with co-detections of RSV subtypes A and B are not well defined. Each genotype tends to dominate for a few consecutive seasons before being displaced by the other genotype, and the predominating genotype and length of circulation changes geographically [25,26]. Therefore, this shift to RSV-B predominance is not unexpected.

4.2. Analysis of the G-Gene Identifies Several Amino Acid Substitutions within RSV-A and RSV-B Samples Circulating in the JHHS

Analysis of the G-gene for RSV-A revealed 12 prominent AASs that were found across all or most of the six clades identified within the GA2.3.5 samples. All 12 AASs were found in the extracellular domain of the G-protein, within two highly glycosylated mucin-like regions (aa66-160 and aa $\approx$ 192–319) that are known to be highly variable [27,28]. Three mutations detected in our cohort, I134K (98.1%), S243I (100%), and K262E (98.1%), are most likely reversion mutations, often described to occur in RSV to change antigenicity, as older ON1 strains contain K134I, I243S, and E262K substitutions [1,26,29,30]. The substitutions I243S, E262K, and K134I were observed in the 2014–2015 season in China and the 2016–2017 season in Taiwan and subsequently detected in multiple countries such as Lebanon, Iran, and Italy, indicating prolonged global circulation [1,26,29–32]. An E224G was detected in an Iranian cohort during the 2018–2019 season in an epitope in escape mutants [31]. An E224V substitution was observed to have emerged in an Italian cohort, further supporting that this site is prone to substitutions resulting from immune pressure [1]. A study showed that glutamic acid (E) at position 262 was under positive selective pressure [30,33]. In our cohort, a G224E (69.2%) substitution was observed, which might also lead to antibody escape as described above. A T319I substitution was described to be circulating in Iran during the 2016–2017 season and subsequently found in 100% of samples in Lebanon in 2019, indicating global circulation of this substitution [31,34]. The Y304H and T320A substitutions, found in 82.7% and 73.1% of our samples, have been observed to be common among ON1 lineages and were the most common AASs detected in Taiwan, Lebanon, Germany, Senegal, China, Iran, and Italy [1,24,26,29,31]. Taiwanese, Lebanese, and Iranian studies reported substitutions, with differing amino acid identities, at position 320, which resulted in the loss of an N-glycosylation site [29,32,34]. In one study, the loss of this glycosylation site was identified as one of several changes thought to be responsible for a change in disease severity in Rome, Italy, during the 2016–2018 seasons [32].

Within RSV-B, seven AASs were detected across the three clades within the GB5.0.5a G-clade. All seven AASs were found in the extracellular domain of the G-protein, within two highly glycosylated mucin-like regions (aa66-160 and aa $\approx$ 192–319) that are known to be highly variable [27,28]. One AAS, A74V (100%), detected in our cohort was previously identified in genomes from pediatric patients in Senegal [35]. Multiple studies detected an A131T amino acid substitution within the 2016–2022 seasons across Asia and Europe, with different countries reporting its presence across various seasons [1,26]. A T137I substitution was described to often be found with the A131T substitution as well as two others, T288I and T310I, in a Chinese study during the 2016–2019 seasons [26]. In our cohort, there was a T131A (96.1%) substitution and an I137T substitution (97%), indicating a reversion back to an older strain, which has been documented within RSV to change antigenic properties to evade host antibodies [36]. A 15-year study out of Kilifi, Kenya, observed R137K/T during the 2007–2008 season where RSV-B predominated, indicating that this site has undergone past changes. In a proposal for RSV nomenclature, 137I was determined to be a defining mutation for RSV-B genotype GB1 [37]. I288T and I310T were found in only one sample, which also contained T131A and I137T. This illustrates that RSV substitutions are spatially and temporally dynamic. An S100G emerged within the 2022–2023 surveillance season in Italy, northwest Spain, and Bulgaria [1,22]. Our samples also displayed the S100G (98.5%) substitution, indicating a potential fitness advantage and global spread. This substitution has become prominent within RSV-B and is a clade-defining mutation for B5.0.5a in a novel RSV nomenclature proposal by Goya et al. [37].

4.3. Analysis of the F-Gene Identifies Several Amino Acid Substitutions within RSV-A and RSV-B Samples Circulating in the JHHS

The RSV F-protein is essential in the RSV viral life cycle and, along with the G-protein, promotes membrane fusion and viral entry [2,38]. It is synthesized as an inactive precursor protein, which is subsequently cleaved at two sites, aa109 and aa136, into two subunits, F1 (137-57) and F2 (aa26-109) [38]. There are six main antigenic sites within the F-protein: Ø (aa92-96, aa195-227), I (aa27-45, aa312-318, aa379-389), II (aa254-277), III (aa46-54, aa301-311, aa345-353, aa367-378), IV (aa422-471), and V (aa55-61, aa146-194, aa287-300) [23]. Similarly, there are some important cytotoxic T-lymphocyte epitopes within the fusion protein [39]. The pre- and post-cleavage forms of the F-protein have distinct as well as shared antigenic sites. Site Ø and V, which are considered the most immunogenic, are found on the pre-fusion conformation whereas sites I, II, III, and IV are shared [40]. Although the F-protein is highly conserved, the signal peptide, transmembrane domain, and antigenic site Ø have been observed to be the most variable [39].

Within RSV-A, two positions were found to have AASs across most samples: T122A (43.4%) substitution and two AASs at position 127, V127I (24.5%) and V127A (9.4%). Both positions are within the p27 portion of the F-protein. A global study from the RSV 2017–2018 seasons detected a T122A substitution within the RSV-A F-protein, indicating that within the conserved F-protein, this site has undergone changes over time [23]. Similarly, a study conducted in China and the Americas found an A122T substitution in a p27 B-cell epitope, but its impact was not fully described [26,41]. A South African study found a reversion substitution A122T paired with a known escape substitution at site 123, suggesting these amino acids may be essential for maintaining protein functionality [39]. Further studies should investigate whether the substitution at position 122 affects F-protein function. Studies from Washington State in 2022–2023 and Texas in 2017 also detected a T122A substitution, suggesting its circulation in the United States in previous seasons [13,41]. Position 122 also falls within a cytotoxic T-lymphocyte epitope [28,39].

Within RSV-B two AASs were identified: S211N, found in the antigenic site Ø at the top of the protein, was present in 96.2% of our RSV-B cohort [42]. Position 389 has substitutions S389P (93%) and S389L (4.2%), which are located in antigenic site I. The S211N substitution is within the nirsevimab binding domain [20]. This substitution has been extensively studied and has no known effect on the neutralization activity of nirsevimab. The S211N substitution has become increasingly common since 2020. Substitutions at position 389 have been detected in South Africa, Brazil, and Bulgaria, indicating that this site is prone to changes [23]. Palivizumab binds within antigenic site I, where these substitutions are located. However, it has been reported that these AASs are not predicted to impact palivizumab binding [23].

Although RSV causes a significant disease burden, there exist only a few treatment and prevention options [43]. All of these target the F-protein, which exhibits high sequence and antigenic conservation between genotypes (>90% identity), is essential in the viral life cycle, and produces high levels of cross-reactive neutralizing antibodies during natural infection, making it a valuable target [43,44]. The use of mAb therapies can increase selective pressure and facilitate antibody escape mutations [43]. To date, few AASs have been detected that impact monoclonal antibody treatments [12,43]. Only one AAS, S276N, was detected in three RSV-A samples and is known to impact palivizumab binding [12]. This substitution has been detected in South Africa, Korea, China, Iran, and Lebanon [12].

Additionally, N201S and Q209L substitutions have been observed in South Africa, the Netherlands, Korea, Brazil, and the USA, affecting nirsevimab binding [12]. Four samples contained an N201K substitution, and eight contained R209Q/K substitutions. Although these positions are known to affect antibody binding, the implications of specific amino acid identities should be further investigated. Therefore, utilizing surveillance systems to monitor and detect potential escape mutations is critical for informing public health and clinical interventions.

5. Conclusions

RSV is the leading cause of acute lower respiratory tract infection in children worldwide and represents a significant disease burden within the United States. This study describes the 2023/2024 RSV season in the Johns Hopkins Health System. Samples from 406 patients were collected for RSV genomic characterization, resulting in 384 sequences. The RSV-B GB5.0.5a clade was predominantly detected, indicating a shift from previous RSV-A dominated seasons within the United States. The RSV-A GA2.3.5 clade was identified to a lesser extent within our cohort. Despite RSV-A circulating to a lesser extent, patients with RSV-A detections were associated with higher admission rates. Multiple amino acid substitutions were detected within both the RSV-A and RSV-B G-protein that have been associated with changes in antigenicity. Further in vitro studies are necessary to describe the impact of each AAS on antigenicity and their potential impact on monoclonal antibody function. Similarly, when the F-gene was examined, amino acid substitutions were detected within both RSV-A and RSV-B. Some of these substitutions or their locations have previously been described to impact the binding of monoclonal antibodies. Following the approval of AREXVY, a prefusion F-protein subunit vaccine against RSV, genomic surveillance will be critical for monitoring escape mutants. Given the diversity of RSV populations within a country, local surveillance efforts are crucial for monitoring RSV activity and genomic evolution.

Supplementary Materials: The following supporting information can be downloaded at https://www.mdpi.com/article/10.3390/v16071122/s1, Table S1: Reference sequences used for the phylogenetic analysis for RSV-A and RSV-B. Table S2: Patient metadata stratified by age group. Supplementary datasets 1–4: fasta files of RSV-A G, RSV-B G, RSV-A F, and RSV-B F-genes.

Author Contributions: Conceptualization, M.Y. and H.H.M.; methodology, M.Y., J.M.N., D.A.V., A.F., O.A. and H.H.M.; software, M.Y., A.F., E.K. and H.H.M.; validation, M.Y., A.P., E.K. and H.H.M.; formal analysis, M.Y. and A.F.; investigation, M.Y. and H.H.M.; resources, A.P. and H.H.M.; data curation, M.Y., E.K. and H.H.M.; writing—original draft preparation, M.Y.; writing—review and editing, M.Y., J.M.N., A.P., A.F. and H.H.M.; visualization, A.P., E.K. and H.H.M.; supervision, H.H.M.; project administration, A.P. and H.H.M.; funding acquisition, A.P. and H.H.M. All authors have read and agreed to the published version of the manuscript.

Funding: This research received no external funding.

Institutional Review Board Statement: The study was conducted in accordance with the Declaration of Helsinki and approved by the Institutional Review Board of Johns Hopkins School of Medicine. The research was performed under protocols IRB00306448 (approval date: 8 November 2022) and IRB00331396 (approval date: 29 October 2019).

Informed Consent Statement: Patient consent was waived as the study was performed using discard samples after the standard-of-care diagnostic testing.

Data Availability Statement: G and F fasta files are attached as Supplementary datasets 1–4.

Acknowledgments: We thank the JHHS medical microbiology laboratories for their assistance with this study.

Conflicts of Interest: The authors declare no relevant conflicts of interest.

References

1. Tramuto, F.; Maida, C.M.; Mazzucco, W.; Costantino, C.; Amodio, E.; Sferlazza, G.; Previti, A.; Immordino, P.; Vitale, F. Molecular Epidemiology and Genetic Diversity of Human Respiratory Syncytial Virus in Sicily during Pre- and Post-COVID-19 Surveillance Seasons. *Pathogens* **2023**, *12*, 1099. [CrossRef] [PubMed]
2. Korsun, N.; Trifonova, I.; Madzharova, I.; Alexiev, I.; Uzunova, I.; Ivanov, I.; Velikov, P.; Tcherveniakova, T.; Christova, I. Resurgence of respiratory syncytial virus with dominance of RSV-B during the 2022–2023 season. *Front. Microbiol.* **2024**, *15*, 1376389. [CrossRef] [PubMed]
3. Staadegaard, L.; Caini, S.; Wangchuk, S.; Thapa, B.; de Almeida, W.A.F.; de Carvalho, F.C.; Njouom, R.; Fasce, R.A.; Bustos, P.; Kyncl, J.; et al. The Global Epidemiology of RSV in Community and Hospitalized Care: Findings From 15 Countries. *Open Forum Infect. Dis.* **2021**, *8*, ofab159. [CrossRef] [PubMed]
4. Obando-Pacheco, P.; Justicia-Grande, A.J.; Rivero-Calle, I.; Rodríguez-Tenreiro, C.; Sly, P.; Ramilo, O.; Mejías, A.; Baraldi, E.; Papadopoulos, N.G.; Nair, H.; et al. Respiratory Syncytial Virus Seasonality: A Global Overview. *J. Infect. Dis.* **2018**, *217*, 1356–1364. [CrossRef] [PubMed]
5. Staadegaard, L.; Caini, S.; Wangchuk, S.; Thapa, B.; de Almeida, W.A.F.; de Carvalho, F.C.; Fasce, R.A.; Bustos, P.; Kyncl, J.; Novakova, L.; et al. Defining the seasonality of respiratory syncytial virus around the world: National and subnational surveillance data from 12 countries. *Influenza Other Respir. Viruses* **2021**, *15*, 732–741. [CrossRef] [PubMed]
6. McGuiness, C.B.; Boron, M.L.; Saunders, B.; Edelman, L.; Kumar, V.R.; Rabon-Stith, K.M. Respiratory syncytial virus surveillance in the United States, 2007–2012: Results from a national surveillance system. *Pediatr. Infect. Dis. J.* **2014**, *33*, 589–594. [CrossRef] [PubMed]
7. McGuiness, C.B.; Boron, M.L.; Saunders, B.; Edelman, L.; Kumar, V.R.; Rabon-Stith, K.M. Molecular Diversity of Human Respiratory Syncytial Virus before and during the COVID-19 Pandemic in Two Neighboring Japanese Cities. *Microbiol. Spectr.* **2023**, *11*, e0260622.
8. Jarrett, J.; Uhteg, K.; Forman, M.S.; Hanlon, A.; Vargas, C.; Carroll, K.C.; Valsamakis, A.; Mostafa, H.H. Clinical performance of the GenMark Dx ePlex respiratory pathogen panels for upper and lower respiratory tract infections. *J. Clin. Virol.* **2021**, *135*, 104737. [CrossRef]
9. Mostafa, H.H.; Carroll, K.C.; Hicken, R.; Berry, G.J.; Manji, R.; Smith, E.; Rakeman, J.L.; Fowler, R.C.; Leelawong, M.; Butler-Wu, S.M.; et al. Multi-center Evaluation of the Cepheid Xpert(R) Xpress SARS-CoV-2/Flu/RSV Test. *J. Clin. Microbiol.* **2020**, *59*, 10–128.
10. Fall, A.; Eldesouki, R.E.; Sachithanandham, J.; Morris, C.P.; Norton, J.M.; Gaston, D.C.; Forman, M.; Abdullah, O.; Gallagher, N.; Li, M.; et al. The displacement of the SARS-CoV-2 variant Delta with Omicron: An investigation of hospital admissions and upper respiratory viral loads. *EBioMedicine* **2022**, *79*, 104008. [CrossRef]
11. Wang, L.; Ng, T.F.F.; Castro, C.J.; Marine, R.L.; Magaña, L.C.; Esona, M.; Peret, T.C.T.; Thornburg, N.J. Next-generation sequencing of human respiratory syncytial virus subgroups A and B genomes. *J. Virol. Methods* **2022**, *299*, 114335. [CrossRef]
12. Langedijk, A.C.; Harding, E.R.; Konya, B.; Vrancken, B.; Lebbink, R.J.; Evers, A.; Willemsen, J.; Lemey, P.; Bont, L.J. A systematic review on global RSV genetic data: Identification of knowledge gaps. *Rev. Med. Virol.* **2022**, *32*, e2284. [CrossRef] [PubMed]
13. Goya, S.; Sereewit, J.; Pfalmer, D.; Nguyen, T.V.; Bakhash, S.; Sobolik, E.B.; Greninger, A.L. Genomic Characterization of Respiratory Syncytial Virus during 2022–2023 Outbreak, Washington, USA. *Emerg. Infect. Dis.* **2023**, *29*, 865–868. [CrossRef] [PubMed]
14. Holland, L.A.; Holland, S.C.; Smith, M.F.; Leonard, V.R.; Murugan, V.; Nordstrom, L.; Mulrow, M.; Salgado, R.; White, M.; Lim, E.S. Genomic Sequencing Surveillance to Identify Respiratory Syncytial Virus Mutations, Arizona, USA. *Emerg. Infect. Dis.* **2023**, *29*, 2380–2382. [CrossRef]
15. Adams, G.; Moreno, G.K.; Petros, B.A.; Uddin, R.; Levine, Z.; Kotzen, B.; Messer, K.S.; Dobbins, S.T.; DeRuff, K.C.; Loreth, C.M.; et al. Viral Lineages in the 2022 RSV Surge in the United States. *N. Engl. J. Med.* **2023**, *388*, 1335–1337. [CrossRef]
16. Ruzin, A.; Pastula, S.T.; Levin-Sparenberg, E.; Jiang, X.; Fryzek, J.; Tovchigrechko, A.; Lu, B.; Qi, Y.; Liu, H.; Jin, H.; et al. Characterization of circulating RSV strains among subjects in the OUTSMART-RSV surveillance program during the 2016–2017 winter viral season in the United States. *PLoS ONE* **2018**, *13*, e0200319. [CrossRef]
17. Pierangeli, A.; Nenna, R.; Fracella, M.; Scagnolari, C.; Oliveto, G.; Sorrentino, L.; Frasca, F.; Conti, M.G.; Petrarca, L.; Papoff, P.; et al. Genetic diversity and its impact on disease severity in respiratory syncytial virus subtype-A and -B bronchiolitis before and after pandemic restrictions in Rome. *J. Infect.* **2023**, *87*, 305–314. [CrossRef]
18. Pruccoli, G.; Castagno, E.; Raffaldi, I.; Denina, M.; Barisone, E.; Baroero, L.; Timeus, F.; Rabbone, I.; Monzani, A.; Terragni, G.M.; et al. The Importance of RSV Epidemiological Surveillance: A Multicenter Observational Study of RSV Infection during the COVID-19 Pandemic. *Viruses* **2023**, *15*, 280. [CrossRef] [PubMed]
19. Redlberger-Fritz, M.; Springer, D.N.; Aberle, S.W.; Camp, J.V.; Aberle, J.H. Respiratory syncytial virus surge in 2022 caused by lineages already present before the COVID-19 pandemic. *J. Med. Virol.* **2023**, *95*, e28830. [CrossRef]
20. Walker, G.J.; Foster, C.S.P.; Sevendal, A.; Domazetovska, A.; Kamalakkannan, A.; Williams, P.C.M.; Kim, K.W.; Condylios, A.; Stelzer-Braid, S.; Bartlett, A.W.; et al. Clinical, Genomic, and Immunological Characterization of RSV Surge in Sydney, Australia 2022. *Pediatrics* **2024**, *153*, e2023063667. [CrossRef]
21. Dolores, A.; Stephanie, G.; Mercedes, S.N.; Érica, G.; Mistchenko, A.S.; Mariana, V. RSV reemergence in Argentina since the SARS-CoV-2 pandemic. *J. Clin. Virol.* **2022**, *149*, 105126. [CrossRef] [PubMed]

22. Davina-Nunez, C.; Perez-Castro, S.; Cabrera-Alvargonzalez, J.J.; Montano-Barrientos, J.; Godoy-Diz, M.; Regueiro, B. The Modification of the Illumina(®) CovidSeq™ Workflow for RSV Genomic Surveillance: The Genetic Variability of RSV during the 2022–2023 Season in Northwest Spain. *Int. J. Mol. Sci.* **2023**, *24*, 16055. [CrossRef] [PubMed]
23. Tabor, D.E.; Fernandes, F.; Langedijk, A.C.; Wilkins, D.; Lebbink, R.J.; Tovchigrechko, A.; Ruzin, A.; Kragten-Tabatabaie, L.; Jin, H.; Esser, M.T.; et al. Global Molecular Epidemiology of Respiratory Syncytial Virus from the 2017–2018 INFORM-RSV Study. *J. Clin. Microbiol.* **2020**, *59*, 10–128. [CrossRef] [PubMed]
24. Fall, A.; Elawar, F.; Hodcroft, E.B.; Jallow, M.M.; Toure, C.T.; Barry, M.A.; Kiori, D.E.; Sy, S.; Diaw, Y.; Goudiaby, D.; et al. Genetic diversity and evolutionary dynamics of respiratory syncytial virus over eleven consecutive years of surveillance in Senegal. *Infect. Genet. Evol.* **2021**, *91*, 104864. [CrossRef] [PubMed]
25. Pangesti, K.N.A.; Ansari, H.R.; Bayoumi, A.; Kesson, A.M.; Hill-Cawthorne, G.A.; Abd El Ghany, M. Genomic characterization of respiratory syncytial virus genotypes circulating in the paediatric population of Sydney, NSW, Australia. *Microb. Genom.* **2023**, *9*, 001095. [CrossRef] [PubMed]
26. Sun, Y.P.; Lei, S.Y.; Wang, Y.B.; Wang, Y.Z.; Qiang, H.S.; Yin, Y.F.; Jiang, Z.M.; Zhu, M.; Chen, X.L.; Ye, H.M.; et al. Molecular Evolution of Attachment Glycoprotein (G) and Fusion Protein (F) Genes of Respiratory Syncytial Virus ON1 and BA9 Strains in Xiamen, China. *Microbiol. Spectr.* **2022**, *10*, e0208321. [CrossRef] [PubMed]
27. Anderson, L.J.; Jadhao, S.J.; Paden, C.R.; Tong, S. Functional Features of the Respiratory Syncytial Virus G Protein. *Viruses* **2021**, *13*, 1214. [CrossRef] [PubMed]
28. Ren, L.; Xia, Q.; Xiao, Q.; Zhou, L.; Zang, N.; Long, X.; Xie, X.; Deng, Y.; Wang, L.; Fu, Z.; et al. The genetic variability of glycoproteins among respiratory syncytial virus subtype A in China between 2009 and 2013. *Infect. Genet. Evol.* **2014**, *27*, 339–347. [CrossRef] [PubMed]
29. Lee, C.Y.; Fang, Y.P.; Wang, L.C.; Chou, T.Y.; Liu, H.F. Genetic Diversity and Molecular Epidemiology of Circulating Respiratory Syncytial Virus in Central Taiwan, 2008–2017. *Viruses* **2021**, *14*, 32. [CrossRef]
30. Song, J.; Zhu, Z.; Song, J.; Mao, N.; Cui, A.; Xu, W.; Zhang, Y. Circulation pattern and genetic variation of human respiratory syncytial virus in China during 2008–2021. *J. Med. Virol.* **2023**, *95*, e28611. [CrossRef]
31. Tavakoli, F.; Izadi, A.; Yavarian, J.; Sharifi-Zarchi, A.; Salimi, V.; Mokhtari-Azad, T. Determination of genetic characterization and circulation pattern of Respiratory Syncytial Virus (RSV) in children with a respiratory infection, Tehran, Iran, during 2018–2019. *Virus Res.* **2021**, *305*, 198564. [CrossRef] [PubMed]
32. Midulla, F.; Di Mattia, G.; Nenna, R.; Scagnolari, C.; Viscido, A.; Oliveto, G.; Petrarca, L.; Frassanito, A.; Arima, S.; Antonelli, G.; et al. Novel Variants of Respiratory Syncytial Virus A ON1 Associated With Increased Clinical Severity of Bronchiolitis. *J. Infect. Dis.* **2020**, *222*, 102–110. [CrossRef]
33. Yu, J.M.; Fu, Y.H.; Peng, X.L.; Zheng, Y.P.; He, J.S. Genetic diversity and molecular evolution of human respiratory syncytial virus A and B. *Sci. Rep.* **2021**, *11*, 12941. [CrossRef] [PubMed]
34. Abou-El-Hassan, H.; Massaad, E.; Soudani, N.; Assaf-Casals, A.; Shaker, R.; Lteif Khoury, M.; Ghanem, S.; Karam, M.; Andary, R.; Saito, R.; et al. Detection of ON1 and novel genotypes of human respiratory syncytial virus and emergence of palivizumab resistance in Lebanon. *PLoS ONE* **2019**, *14*, e0212687. [CrossRef] [PubMed]
35. Jallow, M.M.; Diagne, M.M.; Sagne, S.N.; Tall, F.; Diouf, J.B.N.; Boiro, D.; Mendy, M.P.; Ndiaye, N.K.; Kiori, D.; Sy, S. Respiratory syncytial virus in pediatric patients with severe acute respiratory infections in Senegal: Findings from the 2022 sentinel surveillance season. *Sci. Rep.* **2023**, *13*, 20404. [CrossRef]
36. Otieno, J.R.; Agoti, C.N.; Gitahi, C.W.; Bett, A.; Ngama, M.; Medley, G.F.; Cane, P.A.; Nokes, D.J. Molecular Evolutionary Dynamics of Respiratory Syncytial Virus Group A in Recurrent Epidemics in Coastal Kenya. *J. Virol.* **2016**, *90*, 4990–5002. [CrossRef] [PubMed]
37. Goya, S.; Galiano, M.; Nauwelaers, I.; Trento, A.; Openshaw, P.J.; Mistchenko, A.S.; Zambon, M.; Viegas, M. Toward unified molecular surveillance of RSV: A proposal for genotype definition. *Influenza Other Respir. Viruses* **2020**, *14*, 274–285. [CrossRef] [PubMed]
38. Rezende, W.; Neal, H.E.; Dutch, R.E.; Piedra, P.A. The RSV F p27 peptide: Current knowledge, important questions. *Front. Microbiol.* **2023**, *14*, 1219846. [CrossRef] [PubMed]
39. Mabilo, P.; Mthiyane, H.; Simane, A.; Subramoney, K.; Treurnicht, F.K. Characterisation of RSV Fusion Proteins from South African Patients with RSV Disease, 2019 to 2020. *Viruses* **2022**, *14*, 2321. [CrossRef]
40. Andreano, E.; Paciello, I.; Bardelli, M.; Tavarini, S.; Sammicheli, C.; Frigimelica, E.; Guidotti, S.; Torricelli, G.; Biancucci, M.; D'Oro, U.; et al. The respiratory syncytial virus (RSV) prefusion F-protein functional antibody repertoire in adult healthy donors. *EMBO Mol. Med.* **2021**, *13*, e14035. [CrossRef]
41. Hause, A.M.; Henke, D.M.; Avadhanula, V.; Shaw, C.A.; Tapia, L.I.; Piedra, P.A. Sequence variability of the respiratory syncytial virus (RSV) fusion gene among contemporary and historical genotypes of RSV/A and RSV/B. *PLoS ONE* **2017**, *12*, e0175792.
42. Saito, M.; Tsukagoshi, H.; Sada, M.; Sunagawa, S.; Shirai, T.; Okayama, K.; Sugai, T.; Tsugawa, T.; Hayashi, Y.; Ryo, A.; et al. Detailed Evolutionary Analyses of the F Gene in the Respiratory Syncytial Virus Subgroup A. *Viruses* **2021**, *13*, 2525. [CrossRef] [PubMed]

43. Beach, S.S.; Hull, M.A.; Ytreberg, F.M.; Patel, J.S.; Miura, T.A. Molecular Modeling Predicts Novel Antibody Escape Mutations in the Respiratory Syncytial Virus Fusion Glycoprotein. *J. Virol.* **2022**, *96*, e0035322. [CrossRef] [PubMed]
44. Wu, S.J.; Schmidt, A.; Beil, E.J.; Day, N.D.; Branigan, P.J.; Liu, C.; Gutshall, L.L.; Palomo, C.; Furze, J.; Taylor, G.; et al. Characterization of the epitope for anti-human respiratory syncytial virus F protein monoclonal antibody 101F using synthetic peptides and genetic approaches. *J. Gen. Virol.* **2007**, *88 Pt 10*, 2719–2723. [CrossRef]

viruses

MDPI

Review

From Forgotten Pathogen to Target for New Vaccines: What Clinicians Need to Know about Respiratory Syncytial Virus Infection in Older Adults

Matteo Boattini [1,2,3,*], André Almeida [4,5], Sara Comini [6], Gabriele Bianco [1,2,7], Rossana Cavallo [1,2] and Cristina Costa [1,2]

1 Microbiology and Virology Unit, University Hospital Città della Salute e della Scienza di Torino, 10126 Turin, Italy; gabrielebnc87@gmail.com (G.B.)
2 Department of Public Health and Paediatrics, University of Torino, 10126 Turin, Italy
3 Lisbon Academic Medical Centre, 1649-028 Lisbon, Portugal
4 Department of Internal Medicine 4, Centro Hospitalar Universitário de Lisboa Central, Centro Clínico Académico de Lisboa, 1169-024 Lisbon, Portugal; andre.almeida@chlc.min-saude.pt
5 NOVA Medical School, Universidade Nova de Lisboa, Centro Clínico Académico de Lisboa, 1169-056 Lisbon, Portugal
6 Operative Unit of Clinical Pathology, Carlo Urbani Hospital, 60035 Jesi, Italy
7 Department of Experimental Medicine, University of Salento, Via Provinciale Monteroni n. 165, 73100 Lecce, Italy
* Correspondence: matteo.boattini@unito.it

Abstract: Respiratory syncytial virus (RSV) is increasingly recognized as being implicated in acute illness in older adults, with a significant weight in hospitalizations for respiratory illness and death. By means of a best-evidence review, this paper aims to investigate whether RSV can be considered a forgotten pathogen in older patients, looking at trends in the literature volume and exploring possible epidemiological and clinical features underlying the focus given to it. We then present an assessment of its disease burden and present and future strategies for its reduction, particularly in light of the recent availability of new vaccines.

Keywords: respiratory syncytial virus; RSV; older adults; elderly; vaccine; pneumonia

Citation: Boattini, M.; Almeida, A.; Comini, S.; Bianco, G.; Cavallo, R.; Costa, C. From Forgotten Pathogen to Target for New Vaccines: What Clinicians Need to Know about Respiratory Syncytial Virus Infection in Older Adults. *Viruses* **2024**, *16*, 531. https://doi.org/10.3390/v16040531

Academic Editor: Donald Seto

Received: 7 March 2024
Revised: 24 March 2024
Accepted: 26 March 2024
Published: 29 March 2024

1. Introduction

Respiratory syncytial virus (RSV) is a well-recognized cause of acute respiratory tract illness among infants, with the first case of RSV-related bronchiolitis described as early as 1957 [1]. However, RSV is also increasingly recognized as being implicated in acute illness in older adults [2–15], contributing to a considerable burden in hospitalizations for respiratory illness while presenting mortality rates that are close to those associated with illness caused by influenza viruses [16–18].

RSV is transmitted through contact between viral particles from infected individuals and oral, nasal, and conjunctival mucous membranes. This contact can happen through the emission of droplets by direct contact or by self-inoculation after touching contaminated surfaces, where the virions can remain viable for several hours [19]. Direct contact is, however, the most common route of transmission [20–22]. RSV typically causes seasonal epidemics worldwide. In temperate climates, these usually occur in winter, whereas in tropical and semitropical climates, seasonal epidemics are usually associated with rainy seasons, and in some contexts, RSV circulation may be documented in as many as eight months of the year [23]. However, these seasonal variations have been disrupted by the COVID-19 pandemics and its containment measures [24]. At present, it is therefore not possible to state with absolute certainty that the circulation of RSV will resume in a manner similar to the pre-pandemic period [25].

In adults, the severity of the acute RSV illness is exacerbated by the presence of respiratory and cardiac comorbidities [10,13,26–38] as well as immunosuppression associated with bone marrow or solid organ transplantation [30,39]. Evidence shows that, after the acute respiratory illness resolves, the virus interferes with the immune system's ability to establish memory, allowing for recurrent infections, particularly reinfections in the same winter season [40]. Therefore, RSV infection incidence is heterogeneous given RSV infection results in an incomplete natural immunity that predisposes hosts to reinfection and limits the implementation of serological studies to define the disease burden [23]. In the last decade, a growing body of literature has begun to point out that RSV is the second pathogen after *Streptococcus pneumoniae* more often implicated globally in deaths of people of all ages with a lower respiratory tract infection as well as causing a burden of disease comparable to that of influenza among people over 70 years of age [41]. The increased clinical focus has been accompanied by considerable interest and efforts in the development of new therapies against RSV, with two vaccines recently approved for active immunization of people over 60 years of age [42,43].

2. Research Questions

In this work, we aimed to provide evidence and comment on the following questions:

(a) Is RSV a forgotten pathogen among older adults? Is there a shift in trend in the importance that is being given to it?
(b) Does RSV disease among older adults present a considerable burden?
(c) What new and developing strategies are there to reduce the need for medical care in older adults? Are these relevant?

3. Materials and Methods

We conducted a best-evidence review by searching all peer-reviewed journals focusing on RSV lower respiratory tract infections among older adults, emphasizing articles published since 2000. PubMed/MEDLINE and Cochrane database were scrutinized for articles using the following key word combinations: "respiratory syncytial virus" and "adults", "respiratory syncytial virus" and "elderly", "respiratory syncytial virus" and "older adults", "respiratory syncytial virus" and "vaccine". After providing an overview on RSV's microbiology and epidemiology, we compared the RSV and influenza virus publication volumes in order to make a rough assessment of the clinical interest in the respective topics. We included clinical studies, reports, reviews, and meta-analyses. We excluded protocols, case reports, and case series as well as studies that were not specific for patients over 60 years of age and those that did not have RSV lower respiratory tract infections as its main subject. We then retraced the course of the literature on RSV in older adults. We identified the evidence in support of RSV as a forgotten pathogen, outlining its aspects and inconsistencies from a clinical, diagnostic, and surveillance point of view. Finally, we illustrated the pipeline of vaccines and new therapeutics.

4. RSV Microbiological and Epidemiological Overview

RSV is an enveloped RNA virus and a member of the Paramyxoviridae family and Pneumovirinae subfamily. Virions are characterized by dimensions of 150 nm or more, usually spherical and made of negative single-stranded ribonucleic acid (3′–5′). This molecular characteristic means that in order to be infectious, it needs to be accompanied by polymerase to generate a chain capable of being translated into protein elements, which is why it is essential for the nucleocapsid to be internalized when it enters the host cell [23,44]. RSV presents minimal antigenic heterogeneity with nucleocapsid proteins (N, P, L e M2-1), envelope and non-structural proteins (M, NS1 e NS2), and transmembrane proteins (F, G e SH) [45]. The role of the G protein in attachment to the host cell and the role of the F protein in penetration into the host cell, as well as fusion between viral and host cell membranes, should be emphasized. This ability is responsible for the formation of syncytia, multi-nucleated cell clusters. Both the G and F proteins are important antigenic targets for immune neutralization. Variability in the F protein correlates with disease severity and effectiveness of monoclonal drug and vaccine development [46]. There are two subtypes of RSV described, A and B, with significant genomic variation between the two that are simultaneously present in most outbreaks, A subtypes typically causing the more severe disease. The phenotypic differences are expressed above all in terms of the antigenic expression of the G protein [47].

As far as older adults are concerned, data from US health systems and epidemiological surveillance suggest that RSV is responsible for up to 10,000 deaths per year in people over 64 as well as 60,000 to 120,000 hospitalizations in this population [48]. Data from the city of London from 1994 to 2011 suggest an incidence of 0.7 hospitalizations per 1000 inhabitants [49]; in New York from 2017 to 2020, the data suggest an incidence of 0.4 to 0.6 per 1000 [5]. The only study available from a tropical context, conducted in Guatemala and based on surveillance data, suggests an incidence of 0.3 per 1000 per year [50].

5. Literature Overview Output

The volume of publications on RSV or influenza viruses in older adults over the last 23 years are reported in Figure 1.

Keywords	Respiratory syncytial virus	Influenza
Adults	2637	27,219
Elderly	2907	22,825
Older adults	2482	22,304
Vaccine	3959	44,227

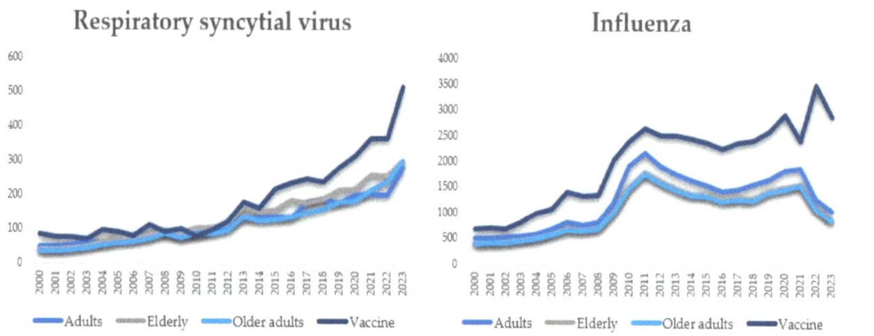

Figure 1. Comparison of respiratory syncytial virus and influenza virus publication volumes (2000–2023).

As a main paper topic, RSV among adults/elderly/older adults was given 7.9 (RSV OR influenza AND elderly) to 11.2 (RSV OR influenza AND vaccine) times less publication volume than its counterparts in influenza. At the same time, it was noted that attention to influenza has always been relatively constant while that for RSV has been on the rise. We finally found a total of 1452 articles that matched our key words, from which we ultimately included a total of 86 for the purposes of this review.

6. Can We Consider RSV a Forgotten Pathogen in Older Adults?

Although "forgotten pathogen" is a formulation that is found every now and then in the scientific literature, especially when referring to infrequent diseases, it does not pertain to a rigorous standardized definition. A search for the term on PubMed/MEDLINE yields just 14 results, with the main suspects being *Streptococcus pneumoniae-* and *Chlamydophila psittaci*-related infections and no results for RSV. Similarly, a search for the term "forgotten disease" provided 287 results with the majority of published papers referring to cases of Lemierre's syndrome. Therefore, these terms seem to be used as narrative artifices to draw the reader's attention rather than to define a clinical underestimation of a given infectious agent in a given population. Although it does not outline the object, the absence of a definition of forgotten pathogen draws attention to the actors involved in forgetting. In the case of RSV in older adults, clinicians, lab diagnostics, surveillance system, and drug companies could have been involved in this process.

6.1. Literature Focus of RSV Infection in Older Adults

The scientific literature focus on RSV in older adults is quite recent. Falsey et al. were among the first to take an interest in the topic with some of their works on factors associated with disease severity and mortality remaining the most common references in the field [2,3]. In the last four years, several systematic reviews and meta-analyses were published [15–17,25,27,29,31,32,41], most of them with the aim to define the disease burden of RSV in older adults and some of them with the active participation of pharma companies involved in RSV vaccine and drug development. Several clinical studies have been published over the years worldwide [51–81]. In Europe, what emerged together with the findings of few multicenter clinical studies [21,59–65] was the need to harmonize and collaborate between the various laboratories in order to obtain shared diagnostic protocols with which to compare data [82].

6.2. Factors Supporting RSV Being a Forgotten Pathogen in Older Adults

Several factors have been reported to be associated with underestimating the RSV disease burden in patients over 60 years [31,83]. These factors relate mainly to clinical features (e.g., non-specificity of symptoms, lack of an effective treatment, absence of a licensed vaccine), diagnostics aspects (e.g., low sensitivity of the antigen-based tests, high costs of polymerase chain reaction (PCR) diagnostic techniques, and presence of low viral loads in older adults), and surveillance-related aspects (e.g., lack of a clinical case definition, non-specificity of the surveillance system).

6.2.1. Clinical Picture and Disease Management

After replicating in the nasopharynx, RSV infects the small bronchiolar epithelium, then extends to alveolar pneumocytes. Pathologic findings of RSV include the necrosis of epithelial cells, occasional proliferation of the bronchiolar epithelium, and infiltration of inflammatory cells between the vascular structures and small airways. This leads to airway obstruction, air trapping, and increased airway resistance [23,84]. However, the symptomatology is non-specific, with an average incubation time of 4–6 days. An asymptomatic infection, a mild infection with symptoms of an upper respiratory tract infection, and a more severe infection with symptoms of a lower respiratory tract infection that can also lead to decompensation of chronic pulmonary and cardiac diseases can be considered the main spectrum of RSV infection presentation (Figure 2) [7,85,86].

Radiology also does not provide much guidance for a diagnosis, as chest tomography findings are very non-specific, as shown by a recent meta-analysis in which the authors, however, state that healthy adults with an RSV infection would have an increased risk for septal thickening, nodular lesions, and ground glass opacities compared to children [87].

Another clinical element that may have contributed to RSV's oblivion in older adults is that there is no effective therapy other than symptomatic, supportive, and complication treatment. Nebulized or oral ribavirin has been shown to reduce morbidity and mortality in hematopoietic stem cell transplant recipients [39,88] but has not been as effective in the immunocompromised and older adults [30,89]. Treatment with glucocorticoids, albeit widely used, is controversial, there being outdated studies in the literature, such as this one showing that short courses of steroids did not affect viral load or shedding [90] and has been associated with longer hospital stays and secondary infections [11], thus leaving room for new studies in this area.

As multicenter clinical studies on RSV infections in older adults, the scenario is quite fragmented, with mortality rates ranging from 0 to 25.9% (Table 1) [21,51–81].

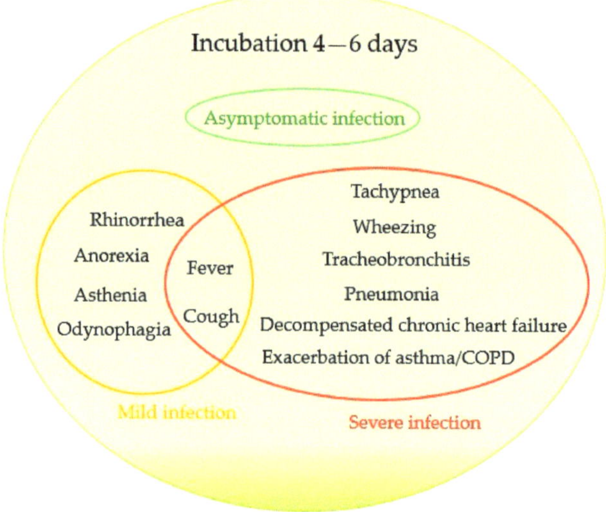

Figure 2. Clinical features of respiratory syncytial virus infection in older adults.

Table 1. Main clinical studies on respiratory syncytial virus infection in older adults according to year of publication.

Study	Period	Country	Patients (n)	Age ± SD or (Range) or [IQR], Years	Critically Ill Patients % (n)	Pneumonia % (n)	Coinfection % (n)	Mortality Rate % (n)
[2]	1999–2003	USA	46 56 132	Not reported Not reported 76 ± 13	Not reported Not reported 15 (20)	2 (1) 7 (4) 31 (41)	Not reported	0 4 (2) 8 (10)
[78]	2007–2008	USA	26 32	65 ± 14 71 ± 13	0 34 (11)	Not reported	Not reported	0 19 (6)
[90]	2005–2008	USA	33 17	69.8 ± 14.9 72.0 ± 14.8	18 (6) 29 (5)	15 (5) 24 (4)	Not reported	6 (2) 0
[79]	2006–2009	USA	31	68 [56–78]	9.7 (3)	Not reported	Not reported	6.5 (2)
[50]	2007–2012	Guatemala	65	≥50	9 (6)	59 (23)	Not reported	13 (8)
[11]	2009–2011	Hong Kong, China	607	75.1 ± 16.4	Not reported	42.3	12.5	9.1
[77]	2009–2010	USA	32	60.8 [44.8–68.9]	16.7 (4)	Not reported	Not reported	4.2 (1)
[51]	2012–2013	Canada	86	74 (19–102)	15 (13)	40 (34)	13 (11)	6 (5)
[4]	2008–2009	15 countries	41	Not reported	Not reported	4.9 (2)	4.9 (2)	Not reported
[72]	2009–2012	USA	41 28 106	53.8 ± 11.8 55 ± 15.1 62.1 ± 19.8	14.6 (6) 17.9 (5) 24.5 (26)	Not reported	Not reported	4.9 (2) 10.7 (3) 6.6 (7)
[69]	2013	Hong Kong, China	123	78 ± 15	12.2 (15)	67.5 (83)	Not reported	8.9 (11)
[81]	2012–2013	USA	75 39	>65 Not reported	Not reported	34.7 38.5	Not reported	4 10.3
[60]	2012–2015	France	53	74 (61–84)	15 (8)	44 (23)	Not reported	8 (4)
[54]	2004–2016	USA	243	≥60	0	9.5 (23)	Not reported	0
[21]	2015–2016	Spain	95	57.7	Not reported	33.6 (32)	Not reported	14.7 (14)
[73]	2000–2013	USA	181	59 (18–87)	13 (24)	Not reported	8 (14)	7 (13)
[70]	2005–2014	Switzerland	107 68	60.5 [48–70.6] 50.8 [37.3–59.4]	29.3 (17)	36.9 (62)	23.4 (25)	19 (11)
[55]	2009–2015	USA	489	60 ± 17	27 (132)	38.8 (190)	8.2 (40)	3.9 (19)
[59]	2005–2018	The Netherlands	192	60.7 [50.8–69.2]	16 (30)	Not reported	42.2 (81)	8 (16)
[56]	2014–2015	Thailand	69	72 [58–81]	36.2 (25)	79.7 (55)	8.7 (6)	15.9 (11)
[91]	2013–2016	France	27 62	70 [56–82] 76 [59–85]	66.7 (18) 21 (13)	100 (27) 100 (62)	100 (27) 0	25.9 (7) 17.7 (11)
[57]	2012–2015	Republic of Korea	132	≥65	25 (33)	56.8 (75)	Not reported	10.6 (14)
[12]	2011–2015	USA	664	78 (60–103)	18	66	Not reported	5.6
[92]	2017–2019	China	113	64.2 ± 16.3	22.1 (25)	Not reported	Not reported	11.5 (13)
[93]	2014–2019	France	616 85	70.4 ± 19.4 66.6 ± 18.6	Not reported	Not reported	0 100 (85)	4.9 (30) 12.9 (11)

Table 1. *Cont.*

Study	Period	Country	Patients (n)	Age ± SD or (Range) or [IQR], Years	Critically Ill Patients % (n)	Pneumonia % (n)	Coinfection % (n)	Mortality Rate % (n)
[68]	2016–2018	Alaska, USA	8	68 [52–77]	Not reported	75(6)	0	0
[76]	2013–2015	USA	192	≥65	13 (25)	Not reported	20.3 (39)	5.9 (11)
[67]	2019–2019	Austria	103	57 [40–73]	6.8 (7)	17.5 (18)	Not reported	2.9 (3)
[37]	2012–2015	New Zealand	281	(18–80)	2.8 (8)	Not reported	Not reported	1.4 (4)
[35]	2017–2019	Belgium, UK, The Netherlands	59	75 (70–79)	0	Not reported	Not reported	0
[62]	2017–2019	Italy, Portugal, Cyprus	166	80.9 ± 8.7	Not reported	29.6 (49)	Not reported	12.1 (20)
[63]	2017–2019	Switzerland	79	78 [65–84]	19 (15)	40.5 (32)	Not reported	10.1 (8)
[71]	2015–2017	USA	1713	≥65 (60%)	20 (344)	Not reported	Not reported	5 (86)
[75]	2016–2018	China	71	77 [67–83]	4.2 (3)	46.5 (33)	21.1 (15)	7 (5)
[65]	2017–2018	Finland	152	73 [65–86]	3.9 (6)	37.5 (57)	Not reported	8.6 (13)
[61]	2011–2018	France, Belgium	309	67.2 ± 15	100 (309)	Not reported	27.2 (84)	23.9 (74)
[74]	2017–2019	USA	403	69.0 [57.2–82.1]	16.4 (66)	Not reported	Not reported	7.7 (31)
[64]	2015–2019	France	1168	75 [63–85]	24.6 (288)	Not reported	18.2 (213)	6.6 (77)
[58]	2016–2019	USA	622	≥65	12.4	Not reported	Not reported	1.5 (9)

Most clinical studies have focused on investigating factors associated with the development of a low respiratory tract infection, especially at the population level, the risk of hospitalization, ICU admission or the need for ventilatory support, and mortality, often compared to a control population (especially patients with influenza virus infections) (Table 2).

Table 2. Factors associated with RSV acute respiratory infection, hospitalization, requirement of ventilatory support, and mortality in older adults.

Study	Acute Lower Respiratory Infection	Hospitalization	Requirement for Ventilatory Support or ICU Admission	Short-Term Mortality
[26]		- Chronic pulmonary disease; - Functional disability; - Low serum neutralizing antibody titre;		
[38]		- Underlying medical conditions; - Female sex; - Increased mucosal IL-6 level; - Longer duration of virus shedding;		
[11]			- Chronic lung disease; - Pneumonia; - Elevated urea and ALT;	- Advanced age; - Pneumonia; - Requirement for ventilation; - Bacterial superinfection; - Elevated urea and WBC count;
[80]	- Congestive heart failure; - Exposure to children;			
[50]			- Cardiovascular disease;	
[51]				- Need for ICU and mechanical ventilation; - Older age; - Major comorbidities;
[69]				- Bacterial superinfection; - Requirement for ventilation;
[72]				- Age > 60 years (vs. age ≤ 60)
[94]				- Lower respiratory infection, chronic respiratory disease, bacterial coinfection, and fever;
[60]	- Cancer - Immunosuppressive treatment;			
[54]		- Age ≥ 75 years (vs. 60–64 years); - COPD or congestive heart failure;		
[73]	- Neutropenia and lymphocytopenia and not receiving ribavirin-based therapy during RSV upper respiratory tract infection;			- Neutropenia and lymphocytopenia at RSV diagnosis;
[70]		- Solid tumours or leukaemia, chronic immunosuppression (vs. HSCT recipients);		

Table 2. *Cont.*

Study	Acute Lower Respiratory Infection	Hospitalization	Requirement for Ventilatory Support or ICU Admission	Short-Term Mortality
[59]				- Lower respiratory tract infection, chronic pulmonary disease, temperature, confusion, and elevated urea;
[8]		- Older age; - COPD; - Congestive heart failure; - Chronic kidney disease; - Previous pneumonia; - Haematological malignancies; - Stroke; - Baseline healthcare resource use;		
[12]				- ≥ two hospitalizations in the prior six months; - Tachypnoea; - Altered consciousness; - Lymphoma; - During hospitalization: - Acute renal failure; - Atrial fibrillation; - Neurovascular complication;
[76]			- Neurologic disease; - Respiratory disease; - Congestive heart failure;	
[62]			- OSA/OHS; - Chronic kidney disease;	- Male gender; - Solid neoplasm; - OSA/OHS;
[67]		- Age > 65 years; - Smoking; - Cardiac disease; - Diabetes mellitus; - Pneumonia;	- Respiratory disease; - Complications; - Pneumonia; - Superinfection;	- Age > 65 years;
[37]		- Age 65–80 and diabetes mellitus; - Age ≥ 50 years and chronic heart failure or COPD;		
[71]		- Higher census tract-level poverty and crowding;		
[75]			- IL-6 concentration;	
[5]	- COPD; - Coronary artery disease; - Congestive heart failure;			
[64]			- Chronic heart or respiratory failure; - Coinfection;	- Age ≥ 85 years; - Neutropenia; - Acute respiratory failure; - Need for ventilation support; - Withdrawing of life-sustaining therapies;
[95]		- COPD or asthma; - Ischemic heart disease; - Stroke; - Diabetes; - Chronic kidney disease;		

Abbreviations: COPD: chronic obstructive pulmonary disease; ALT: alanine aminotransferase; WBC: white blood cells; ICU: intensive care unit; HSCT: hematopoietic stem cell transplant; OSA/OHS: obstructive sleep apnoea or obesity hypoventilation syndrome.

In Europe, a Portuguese, Italian, and Cypriot project that included patients hospitalized due to a lower respiratory tract infection due to influenza and/or RSV during two consecutive winter periods showed that influenza A H1N1 virus was the only microbiological factor associated with in-hospital mortality and a need for invasive mechanical ventilation [96]. In one of the sub-analyses carried out on the population over 85 years of age, however, it emerges that chronic diseases, especially COPD or asthma and chronic kidney disease KDIGO stage 3A or worse, weigh more on outcomes (pneumonia and death) than viral etiology [97]. In another sub-analysis of the same cohort, which included 166 elderly patients with a lower respiratory tract RSV infection, about 30% of the patients had pneumonia, the in-hospital mortality rate was 12.1%, and the factors associated with mortality were the male gender, solid neoplasia and obstructive sleep apnea, and/or obesity hypoventilation syndrome [62]. In a French multicenter study that included more than 1000 adult patients, the overall mortality rate was lower (around 6%), while it was more similar to the overmentioned cohort for ICU patients. The study reported the factors associiated with a need for mechanical ventilation to be chronic cardio-pulmonary diseases [64]. In a Scottish and Danish study, Osei-Yeboah et al. found an increased risk of RSV hospitalization in adults with chronic obstructive pulmonary disease, ischemic heart disease, and chronic kidney disease when compared with those of the overall population [95].

6.2.2. Diagnostic Challenges

Laboratory diagnostic methods for an RSV diagnosis mainly include rapid antigen testing and PCR. Despite their ease of use, antigenic tests for RSV do not have high sensitivity in adult patients [98,99]. Besides not being cheap and gold standard, PCR techniques have different sensitivities, with single-target PCRs having a higher sensitivity than multiplexes [100]. In a recent meta-analysis, the authors conclude that, for improving RSV detection, at least one other respiratory sample should be tested in addition to the RT-PCR test of the nasopharyngeal swab and that diagnostic sensitivity could benefit from testing up to at least 3 respiratory samples [101], increasing the complexity in terms of logistics and costs.

6.2.3. Recent Developments on Surveillance and Case Definition

RSV surveillance was officially introduced by the WHO in 2016 [101], starting to take into account that, in about 50% of older adults, an RSV infection may occur without fever. A case definition for hospital and community was provided [102]. Extended SARI (severe acute respiratory infection) and ARI (acute respiratory infection) were defined, and the case definitions of influenza-like illness and SARI for influenza [103], which required fever as a symptom and on which both RSV surveillance and clinical studies had previously been based, were abandoned. The RSV surveillance system now includes screening for RSV of patients with symptoms compatible with extended SARI in sentinel hospitals [104]. For older adults, the respiratory sample to be tested is sputum. If the test is positive for RSV, typing in A or B and subsequent notification is performed. If the test is negative and the patient presented with fever, testing for influenza is performed, with the subsequent notification of the result [105].

7. Estimating RSV Disease Burden in Older Adults

In recent years, several meta-analyses aimed at estimating the RSV disease burden in adults have been published, most of them highlighting the high burden of the disease in older adults [16,27–29,31,32]. The first estimate was by Falsey et al., which stated that global incidence could be around 6.3 per 1000 people older than 70 years of age or high-risk adults [2]. A more recent meta-analysis estimated that, in 2015, RSV would cause illness in about 1.5 billion people over 50 years of age, that 14.5% would be hospitalized, and that 1.6% would die [28]. Another meta-analysis estimated a case fatality proportion between 8 and 10 percent in elderly and high-risk adult patients, respectively [27]. In Europe, a Northern European, prospective, observational cohort study aiming at assessing the community burden of RSV in older adults aged $\geq$ 60 years showed that RSV had an incidence between 4 to 7.2%, with mostly mild infections, leading to both fewer medical appointments and antibiotic prescriptions in comparison to influenza-associated infections [35]. Conversely, an estimation of hospital admission in Europe showed that RSV causes about 160,000 admissions per year, with about 92% of cases occurring in adults aged $\geq$ 65 years [106].

7.1. Limitations in Estimating RSV Disease Burden in Older Adults

Almost all the works that attempted to estimate the RSV disease burden in older adults reported several limitations. These are basically due to the processing data from studies performed with very different methods and protocols [107]; being based on the case definition of influenza-like illness [108]; lacking the necessary data to stratify by age group; being based on data mainly derived from academic, high-income, in-hospital settings [31,67]; and underestimating the presence of other factors, such as coinfections or complications [15]. To address these limitations, the WHO published a document with the essential guidelines to be followed in estimating the RSV disease burden [109], but robust evidence is lacking.

7.2. Burden of Coinfections

Among older adults with an RSV infection, the presence of another pathogen in the respiratory tract is another factor to be considered in the disease burden analysis [66]. The rates of viral and bacterial coinfection in causing pneumonia range from 10 to 68% [110], making the associated disease severity a wide area needing investigation. Several bacterial and viral respiratory pathogens, such as *Staphylococcus aureus, Pseudomonas aeruginosa, Streptococcus pneumoniae, Haemophilus influenzae, Moraxella catarrhalis, Klebsiella pneumoniae*, human rhinoviruses, influenza A virus, human metapneumovirus, and human parainfluenza viruses, have been reported to be involved [91,92,94,111]. Celante et al. reported a coinfection rate of 18.2% in their cohort of French hospitalized adults with an RSV infection, and coinfection was associated with a need for invasive mechanical ventilation [64]. Godefroy et al. showed that bacterial coinfections may be present in about 12% of older adults with an RSV infection, and these patients showed a higher mortality rate [93].

8. Pipeline of Vaccines and New Therapeutics

RSV vaccine candidates for pediatric, maternal, or older populations use four approaches: live attenuated, protein based (inactivated, subunit, or particle-based), nucleic acid vaccines, and recombinant vectors [112]. There are several candidates in different stages of study and development [113]. As far as the elderly are concerned, in spring 2023, two vaccines were approved by the responsible drug agencies and placed on the market for the active immunization of people over 60 years of age. Specifically, the first one, marketed by GlaxoSmithKline, is a monovalent vaccine and comprises RSV pre-F protein. It showed a vaccine efficacy of 94.1% in phase 3 trial (NCT04886596) against severe RSV-related lower respiratory tract disease and a satisfactory safety profile [114]. The second one, marketed by Pfizer, is a bivalent vaccine comprising subtype A and B of RSV pre-F proteins. The RENOIR phase 3 trial (NCT05035212) showed a vaccine efficacy of preventing an RSV-associated lower respiratory tract illness between 66.7 (at least two signs or symptoms) and 85.7% (at least three signs or symptoms) [115] and no vaccine-related serious adverse events through 12 months post-vaccination [116]. Concerning RSV vaccine candidates in Phase 3, the Moderna candidate, the only nucleic acid vaccine in phase 3, showed 83.7% (95.88% CI, 66.0–92.2) efficacy (NCT05127434) in preventing confirmed lower respiratory tract RSV infections with at least two lower respiratory symptoms with no safety concerns [117]. Of note, there are several phase 1 studies ongoing, assessing the safety and reactogenicity of combined mRNA vaccines in healthy older adults [118]. As far as new therapies are concerned, monoclonal antibodies that seem to be effective in newborns need to be assessed for efficacy and safety in at-risk adults, currently having no role. There are still other drugs that attempt to target the mechanism of virus entry and replication, some of which have shown promising results, such as fusion [119] and nucleoprotein inhibitors [120].

9. Conclusions

Although the term 'forgotten pathogen' is more akin to a narrative gimmick than a scientific term, one can fairly say that publication interest in RSV among older adults has been relatively low for a long time up until the late 2010s. Several reasons associated with the clinical characteristics of the associated infections, diagnostics, and surveillance have been put forward as contributing to this relative lack of concern. The surveillance system itself is very recent, compared to the influenza system, which is over 60 years old, so a full understanding of the phenomenon is still some time away. Moreover, the limitations of published studies make it difficult to estimate the real disease burden of this pathogen in the older and adult population in general. Interest in RSV among older adults has, however, grown in recent years, in parallel with the final stages of vaccine development, even though the associated disease burden has not been well characterized. In fact, in spite of this knowledge gap, it can be said that RSV has not been forgotten by the companies involved in the development of drugs and vaccines since the first attempts to develop a vaccine against RSV for children date back to the 1960s [121], and the first vaccines to be marketed are those for older adults.

Given the trial evidence so far, RSV vaccination should already be considered the only relevant option for patients living with diabetes mellitus, advanced age, chronic organ diseases, frailty, and immunodeficiency and those residing in long-term care facilities or frequently exposed to young children (Figure 3) in order to reduce the risk of a lower tract respiratory disease and need for medical care [122].

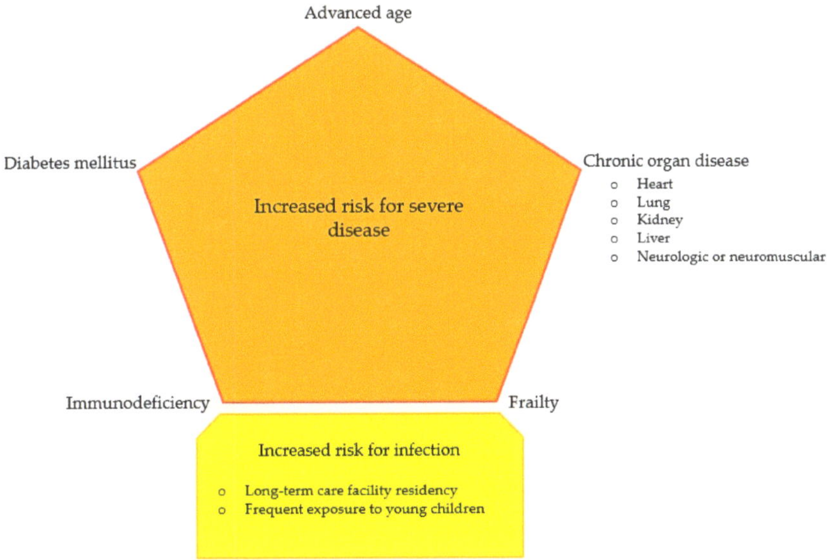

Figure 3. Graphical decision making for respiratory syncytial virus vaccination.

Author Contributions: Conceptualization, M.B. and A.A.; methodology, M.B.; software, G.B.; validation, A.A., S.C. and C.C.; formal analysis, M.B.; resources, M.B.; data curation, M.B.; writing—original draft preparation, M.B.; writing—review and editing, A.A.; visualization, S.C.; supervision, G.B. and R.C.; project administration, R.C. and C.C.; funding acquisition, C.C. All authors have read and agreed to the published version of the manuscript.

Funding: This study was supported by EU funding within the MUR PNRR Extended Partnership initiative on Emerging Infectious Diseases (Project no. PE00000007, INF-ACT).

Data Availability Statement: No new data were created.

Conflicts of Interest: A.A. has received honoraria for teaching lectures sponsored by Pfizer. All other authors declare no conflicts of interest.

References

1. Chanock, R.; Roizman, B.; Myers, R. Recovery from infants with respiratory illness of a virus related to chimpanzee coryza agent (CCA): Isolation, properties and characterization. *Am. J. Epidemiol.* **1957**, *66*, 281–290. [CrossRef] [PubMed]
2. Falsey, A.R.; Hennessey, P.A.; Formica, M.A.; Cox, C.; Walsh, E.E. Respiratory syncytial virus infection in elderly and high-risk adults. *N. Engl. J. Med.* **2005**, *352*, 1749–1759. [CrossRef] [PubMed]
3. Falsey, A.R.; Walsh, E.E. Respiratory syncytial virus infection in elderly adults. *Drugs Aging* **2005**, *22*, 577–587. [CrossRef] [PubMed]
4. Falsey, A.R.; McElhaney, J.E.; Beran, J.; van Essen, G.A.; Duval, X.; Esen, M.; Galtier, F.; Gervais, P.; Hwang, S.-J.; Kremsner, P.; et al. Respiratory syncytial virus and other respiratory viral infections in older adults with moderate to severe influenza-like illness. *J. Infect. Dis.* **2014**, *209*, 1873–1881. [CrossRef] [PubMed]

5. Branche, A.R.; Saiman, L.; Walsh, E.E.; Falsey, A.R.; Sieling, W.D.; Greendyke, W.; Peterson, D.R.; Vargas, C.Y.; Phillips, M.; Finelli, L. Incidence of Respiratory Syncytial Virus Infection among Hospitalized Adults, 2017–2020. *Clin. Infect. Dis.* **2022**, *74*, 1004–1011. [CrossRef] [PubMed]
6. Thompson, W.W.; Shay, D.K.; Weintraub, E.; Brammer, L.; Cox, N.; Anderson, L.J.; Fukuda, K. Mortality associated with influenza and respiratory syncytial virus in the United States. *JAMA* **2003**, *289*, 179–186. [CrossRef] [PubMed]
7. Colosia, A.D.; Yang, J.; Hillson, E.; Mauskopf, J.; Copley-Merriman, C.; Shinde, V.; Stoddard, J. The epidemiology of medically attended respiratory syncytial virus in older adults in the United States: A systematic review. *PLoS ONE* **2017**, *12*, e0182321. [CrossRef] [PubMed]
8. Wyffels, V.; Kariburyo, F.; Gavart, S.; Fleischhackl, R.; Yuce, H. A Real-World Analysis of Patient Characteristics and Predictors of Hospitalization among US Medicare Beneficiaries with Respiratory Syncytial Virus Infection. *Adv. Ther.* **2020**, *37*, 1203–1217. [CrossRef] [PubMed]
9. Childs, A.; Zullo, A.R.; Joyce, N.R.; McConeghy, K.W.; van Aalst, R.; Moyo, P.; Bosco, E.; Mor, V.; Gravenstein, S. The burden of respiratory infections among older adults in long-term care: A systematic review. *BMC Geriatr.* **2019**, *19*, 210. [CrossRef] [PubMed]
10. Han, L.L.; Alexander, J.P.; Anderson, L.J. Respiratory syncytial virus pneumonia among the elderly: An assessment of disease burden. *J. Infect. Dis.* **1999**, *179*, 25–30. [CrossRef] [PubMed]
11. Lee, N.; Lui, G.C.; Wong, K.T.; Li, T.C.; Tse, E.C.; Chan, J.Y.; Yu, J.; Wong, S.S.; Choi, K.W.; Wong, R.Y.; et al. High morbidity and mortality in adults hospitalized for respiratory syncytial virus infections. *Clin. Infect. Dis.* **2013**, *57*, 1069–1077. [CrossRef] [PubMed]
12. Tseng, H.F.; Sy, L.S.; Ackerson, B.; Solano, Z.; Slezak, J.; Luo, Y.; Shinde, V. Severe Morbidity and Short- and Mid- to Long-term Mortality in Older Adults Hospitalized with Respiratory Syncytial Virus Infection. *J. Infect. Dis.* **2020**, *222*, 1298–1310. [CrossRef] [PubMed]
13. Ackerson, B.; Tseng, H.F.; Sy, L.S.; Solano, Z.; Slezak, J.; Luo, Y.; Fischetti, C.A.; Shinde, V. Severe Morbidity and Mortality Associated with Respiratory Syncytial Virus Versus Influenza Infection in Hospitalized Older Adults. *Clin. Infect. Dis.* **2019**, *69*, 197–203. [CrossRef] [PubMed]
14. Fleming, D.M.; Taylor, R.J.; Lustig, R.L.; Schuck-Paim, C.; Haguinet, F.; Webb, D.J.; Logie, J.; Matias, G.; Taylor, S. Modelling estimates of the burden of Respiratory Syncytial virus infection in adults and the elderly in the United Kingdom. *BMC Infect. Dis.* **2015**, *15*, 443. [CrossRef] [PubMed]
15. Domnich, A.; Calabrò, G.E. Epidemiology and burden of respiratory syncytial virus in Italian adults: A systematic review and meta-analysis. *PLoS ONE* **2024**, *19*, e0297608. [CrossRef]
16. Maggi, S.; Veronese, N.; Burgio, M.; Cammarata, G.; Ciuppa, M.E.; Ciriminna, S.; Di Gennaro, F.; Smith, L.; Trott, M.; Dominguez, L.J.; et al. Rate of Hospitalizations and Mortality of Respiratory Syncytial Virus Infection Compared to Influenza in Older People: A Systematic Review and Meta-Analysis. *Vaccines* **2022**, *10*, 2092. [CrossRef] [PubMed]
17. ElSherif, M.; Andrew, M.K.; Ye, L.; Ambrose, A.; Boivin, G.; Bowie, W.; David, M.P.; Gruselle, O.; Halperin, S.A.; Hatchette, T.F.; et al. Leveraging Influenza Virus Surveillance From 2012 to 2015 to Characterize the Burden of Respiratory Syncytial Virus Disease in Canadian Adults ≥50 Years of Age Hospitalized with Acute Respiratory Illness. *Open Forum Infect. Dis.* **2023**, *10*, ofad315. [CrossRef] [PubMed]
18. Ackerson, B.; An, J.; Sy, L.S.; Solano, Z.; Slezak, J.; Tseng, H.F. Cost of Hospitalization Associated with Respiratory Syncytial Virus Infection Versus Influenza Infection in Hospitalized Older Adults. *J. Infect. Dis.* **2020**, *222*, 962–966. [CrossRef] [PubMed]
19. Hall, C.B.; Douglas, R.G., Jr.; Schnabel, K.C.; Geiman, J.M. Infectivity of respiratory syncytial virus by various routes of inoculation. *Infect. Immun.* **1981**, 33, 779–783. [CrossRef] [PubMed]
20. Kulkarni, H.; Smith, C.M.; Lee, D.H.; Hirst, R.A.; Easton, A.J.; O'Callaghan, C. Evidence of Respiratory Syncytial Virus Spread by Aerosol. Time to Revisit Infection Control Strategies? *Am. J. Respir. Crit. Care Med.* **2016**, *194*, 308–316. [CrossRef]
21. Kestler, M.; Muñoz, P.; Mateos, M.; Adrados, D.; Bouza, E. Respiratory syncytial virus burden among adults during flu season: An underestimated pathology. *J. Hosp. Infect.* **2018**, *100*, 463–468. [CrossRef] [PubMed]
22. Korsten, K.; Adriaenssens, N.; Coenen, S.; Butler, C.C.; Pirçon, J.Y.; Verheij, T.J.M.; Bont, L.J.; Wildenbeest, J.G.; RESCEU Investigators. Contact with Young Children Increases the Risk of Respiratory Infection in Older Adults in Europe-the RESCEU Study. *J. Infect. Dis.* **2022**, *226*, S79–S86. [CrossRef] [PubMed]
23. Griffiths, C.; Drews, S.J.; Marchant, D.J. Respiratory Syncytial Virus: Infection, Detection, and New Options for Prevention and Treatment. *Clin. Microbiol. Rev.* **2017**, *30*, 277–319. [CrossRef] [PubMed]
24. Guitart, C.; Bobillo-Perez, S.; Alejandre, C.; Armero, G.; Launes, C.; Cambra, F.J.; Balaguer, M.; Jordan, I.; Hospital Network for R. S. V. surveillance in Catalonia. Bronchiolitis, epidemiological changes during the SARS-CoV-2 pandemic. *BMC Infect. Dis.* **2022**, *22*, 84. [CrossRef]
25. Hamid, S.; Winn, A.; Parikh, R.; Jones, J.M.; McMorrow, M.; Prill, M.M.; Silk, B.J.; Scobie, H.M.; Hall, A.J. Seasonality of Respiratory Syncytial Virus—United States, 2017–2023. *Morb. Mortal. Wkly. Rep. (MMWR)* **2023**, *72*, 355–361. [CrossRef]
26. Walsh, E.E.; Peterson, D.R.; Falsey, A.R. Risk factors for severe respiratory syncytial virus infection in elderly persons. *J. Infect. Dis.* **2004**, *189*, 233–238. [CrossRef] [PubMed]

27. Nguyen-Van-Tam, J.S.; O'Leary, M.; Martin, E.T.; Heijnen, E.; Callendret, B.; Fleischhackl, R.; Comeaux, C.; Tran, T.M.P.; Weber, K. Burden of respiratory syncytial virus infection in older and high-risk adults: A systematic review and meta-analysis of the evidence from developed countries. *Eur. Respir. Rev.* **2022**, *31*, 220105. [CrossRef] [PubMed]
28. Shi, T.; Denouel, A.; Tietjen, A.K.; Campbell, I.; Moran, E.; Li, X.; Campbell, H.; Demont, C.; Nyawanda, B.O.; Chu, H.Y.; et al. Global Disease Burden Estimates of Respiratory Syncytial Virus-Associated Acute Respiratory Infection in Older Adults in 2015: A Systematic Review and Meta-Analysis. *J. Infect. Dis.* **2020**, *222*, S577–S583. [CrossRef] [PubMed]
29. Shi, T.; Vennard, S.; Jasiewicz, F.; Brogden, R.; Nair, H.; RESCEU Investigators. Disease Burden Estimates of Respiratory Syncytial Virus related Acute Respiratory Infections in Adults with Comorbidity: A Systematic Review and Meta-Analysis. *J. Infect. Dis.* **2022**, *226*, S17–S21. [CrossRef] [PubMed]
30. Villanueva, D.H.; Arcega, V.; Rao, M. Review of respiratory syncytial virus infection among older adults and transplant recipients. *Ther. Adv. Infect. Dis.* **2022**, *9*, 20499361221091413. [CrossRef] [PubMed]
31. Savic, M.; Penders, Y.; Shi, T.; Branche, A.; Pirçon, J.Y. Respiratory syncytial virus disease burden in adults aged 60 years and older in high-income countries: A systematic literature review and meta-analysis. *Influenza Other Respir Viruses* **2023**, *17*, e13031. [CrossRef] [PubMed]
32. Tin Tin Htar, M.; Yerramalla, M.S.; Moïsi, J.C.; Swerdlow, D.L. The burden of respiratory syncytial virus in adults: A systematic review and meta-analysis. *Epidemiol. Infect.* **2020**, *148*, e48. [CrossRef] [PubMed]
33. Nam, H.H.; Ison, M.G. Respiratory syncytial virus infection in adults. *BMJ* **2019**, *366*, l5021. [CrossRef] [PubMed]
34. Bouzid, D.; Visseaux, B.; Ferré, V.M.; Peiffer-Smadja, N.; Le Hingrat, Q.; Loubet, P. Respiratory syncytial virus in adults with comorbidities: An update on epidemiology, vaccines, and treatments. *Clin. Microbiol. Infect.* **2023**, *29*, 1538–1550. [CrossRef] [PubMed]
35. Korsten, K.; Adriaenssens, N.; Coenen, S.; Butler, C.; Ravanfar, B.; Rutter, H.; Allen, J.; Falsey, A.; Pirçon, J.Y.; Gruselle, O.; et al. Burden of respiratory syncytial virus infection in community-dwelling older adults in Europe (RESCEU): An international prospective cohort study. *Eur. Respir. J.* **2021**, *57*, 2002688. [CrossRef] [PubMed]
36. Ivey, K.S.; Edwards, K.M.; Talbot, H.K. Respiratory Syncytial Virus and Associations with Cardiovascular Disease in Adults. *J. Am. Coll. Cardiol.* **2018**, *71*, 1574–1583. [CrossRef] [PubMed]
37. Prasad, N.; Walker, T.A.; Waite, B.; Wood, T.; Trenholme, A.A.; Baker, M.G.; McArthur, C.; Wong, C.A.; Grant, C.C.; Huang, Q.S.; et al. Respiratory Syncytial Virus-Associated Hospitalizations among Adults with Chronic Medical Conditions. *Clin. Infect. Dis.* **2021**, *73*, e158–e163. [CrossRef] [PubMed]
38. Walsh, E.E.; Peterson, D.R.; Kalkanoglu, A.E.; Lee, F.E.; Falsey, A.R. Viral shedding and immune responses to respiratory syncytial virus infection in older adults. *J. Infect. Dis.* **2013**, *207*, 1424–1432. [CrossRef]
39. Waghmare, A.; Campbell, A.P.; Xie, H.; Seo, S.; Kuypers, J.; Leisenring, W.; Jerome, K.R.; Englund, J.A.; Boeckh, M. Respiratory syncytial virus lower respiratory disease in hematopoietic cell transplant recipients: Viral RNA detection in blood, antiviral treatment, and clinical outcomes. *Clin. Infect. Dis.* **2013**, *57*, 1731–1741. [CrossRef] [PubMed]
40. Blunck, B.N.; Angelo, L.S.; Henke, D.; Avadhanula, V.; Cusick, M.; Ferlic-Stark, L.; Zechiedrich, L.; Gilbert, B.E.; Piedra, P.A. Adult Memory T Cell Responses to the Respiratory Syncytial Virus Fusion Protein During a Single RSV Season (2018–2019). *Front. Immunol.* **2022**, *13*, 823652. [CrossRef] [PubMed]
41. GBD 2016 Lower Respiratory Infections Collaborators. Estimates of the global, regional, and national morbidity, mortality, and aetiologies of lower respiratory infections in 195 countries, 1990–2016: A systematic analysis for the Global Burden of Disease Study 2016. *Lancet Infect. Dis.* **2018**, *18*, 1191–1210. [CrossRef] [PubMed]
42. European Medicines Agency. Abrysvo. Available online: https://www.ema.europa.eu/en/medicines/human/EPAR/abrysvo (accessed on 20 December 2023).
43. European Medicines Agency. Arexvy. Available online: https://www.ema.europa.eu/en/medicines/human/EPAR/arexvy (accessed on 20 December 2023).
44. International Committee on Taxonomy of Viruses. Family: *Paramysoviridae*. Chapter Version: ICTV Ninth Report; 2009 Taxonomy Release. Available online: https://ictv.global/report_9th/RNAneg/Mononegavirales/Paramyxoviridae (accessed on 24 January 2024).
45. Schobel, S.A.; Stucker, K.M.; Moore, M.L.; Anderson, L.J.; Larkin, E.K.; Shankar, J.; Bera, J.; Puri, V.; Shilts, M.H.; Rosas-Salazar, C.; et al. Respiratory Syncytial Virus whole-genome sequencing identifies convergent evolution of sequence duplication in the C-terminus of the G gene. *Sci. Rep.* **2016**, *6*, 26311. [CrossRef] [PubMed]
46. Ludlow, M. Respiratory syncytial virus infection in the modern era. *Curr. Opin. Infect. Dis.* **2023**, *36*, 155–163. [CrossRef] [PubMed]
47. McLellan, J.S.; Ray, W.C.; Peeples, M.E. Structure and function of respiratory syncytial virus surface glycoproteins. *Curr. Top. Microbiol. Immunol.* **2013**, *372*, 83–104. [CrossRef] [PubMed]
48. CDC. RSV Surveillance & Research. October 2022. Available online: https://www.cdc.gov/rsv/research/index.html (accessed on 29 January 2024).
49. Mangtani, P.; Hajat, S.; Kovats, S.; Wilkinson, P.; Armstrong, B. The association of respiratory syncytial virus infection and influenza with emergency admissions for respiratory disease in London: An analysis of routine surveillance data. *Clin. Infect. Dis.* **2006**, *42*, 640–646. [CrossRef] [PubMed]

50. McCracken, J.P.; Prill, M.M.; Arvelo, W.; Lindblade, K.A.; López, M.R.; Estevez, A.; Müller, M.L.; Muñoz, F.; Bernart, C.; Cortez, M.; et al. Respiratory syncytial virus infection in Guatemala, 2007–2012. *J. Infect. Dis.* **2013**, *208* (Suppl. 3), S197–S206. [CrossRef] [PubMed]
51. Volling, C.; Hassan, K.; Mazzulli, T.; Green, K.; Al-Den, A.; Hunter, P.; Mangat, R.; Ng, J.; McGeer, A. Respiratory syncytial virus infection-associated hospitalization in adults: A retrospective cohort study. *BMC Infect. Dis.* **2014**, *14*, 665. [CrossRef] [PubMed]
52. Sundaram, M.E.; Meece, J.K.; Sifakis, F.; Gasser, R.A., Jr.; Belongia, E.A. Medically attended respiratory syncytial virus infections in adults aged $\geq$ 50 years: Clinical characteristics and outcomes. *Clin. Infect. Dis.* **2014**, *58*, 342–349. [CrossRef] [PubMed]
53. Malosh, R.E.; Martin, E.T.; Callear, A.P.; Petrie, J.G.; Lauring, A.S.; Lamerato, L.; Fry, A.M.; Ferdinands, J.; Flannery, B.; Monto, A.S. Respiratory syncytial virus hospitalization in middle-aged and older adults. *J. Clin. Virol.* **2017**, *96*, 37–43. [CrossRef] [PubMed]
54. Belongia, E.A.; King, J.P.; Kieke, B.A.; Pluta, J.; Al-Hilli, A.; Meece, J.K.; Shinde, V. Clinical Features, Severity, and Incidence of RSV Illness During 12 Consecutive Seasons in a Community Cohort of Adults $\geq$60 Years Old. *Open Forum Infect. Dis.* **2018**, *5*, ofy316. [CrossRef] [PubMed]
55. Schmidt, H.; Das, A.; Nam, H.; Yang, A.; Ison, M.G. Epidemiology and outcomes of hospitalized adults with respiratory syncytial virus: A 6-year retrospective study. *Influenza Other Respir. Viruses* **2019**, *13*, 331–338. [CrossRef] [PubMed]
56. Chuaychoo, B.; Ngamwongwan, S.; Kaewnaphan, B.; Athipanyasilp, N.; Horthongkham, N.; Kantakamalakul, W.; Muangman, N. Clinical manifestations and outcomes of respiratory syncytial virus infection in adult hospitalized patients. *J. Clin. Virol.* **2019**, *117*, 103–108. [CrossRef] [PubMed]
57. Yoon, J.G.; Noh, J.Y.; Choi, W.S.; Park, J.J.; Suh, Y.B.; Song, J.Y.; Cheong, H.J.; Kim, W.J. Clinical characteristics and disease burden of respiratory syncytial virus infection among hospitalized adults. *Sci. Rep.* **2020**, *10*, 12106. [CrossRef] [PubMed]
58. Begley, K.M.; Monto, A.S.; Lamerato, L.E.; Malani, A.N.; Lauring, A.S.; Talbot, H.K.; Gaglani, M.; McNeal, T.; Silveira, F.P.; Zimmerman, R.K.; et al. Prevalence and Clinical Outcomes of Respiratory Syncytial Virus vs Influenza in Adults Hospitalized With Acute Respiratory Illness From a Prospective Multicenter Study. *Clin. Infect. Dis.* **2023**, *76*, 1980–1988. [CrossRef] [PubMed]
59. Vos, L.M.; Oosterheert, J.J.; Hoepelman, A.I.M.; Bont, L.J.; Coenjaerts, F.E.J.; Naaktgeboren, C.A. External validation and update of a prognostic model to predict mortality in hospitalized adults with RSV: A retrospective Dutch cohort study. *J. Med. Virol.* **2019**, *91*, 2117–2124. [CrossRef] [PubMed]
60. Loubet, P.; Lenzi, N.; Valette, M.; Foulongne, V.; Krivine, A.; Houhou, N.; Lagathu, G.; Rogez, S.; Alain, S.; Duval, X.; et al. Clinical characteristics and outcome of respiratory syncytial virus infection among adults hospitalized with influenza-like illness in France. *Clin. Microbiol. Infect.* **2017**, *23*, 253–259. [CrossRef] [PubMed]
61. Coussement, J.; Zuber, B.; Garrigues, E.; Gros, A.; Vandueren, C.; Epaillard, N.; Voiriot, G.; Tandjaoui-Lambiotte, Y.; Lascarrou, J.B.; Boissier, F.; et al. Characteristics and Outcomes of Patients in the ICU with Respiratory Syncytial Virus Compared with Those with Influenza Infection: A Multicenter Matched Cohort Study. *Chest* **2022**, *161*, 1475–1484. [CrossRef] [PubMed]
62. Boattini, M.; Almeida, A.; Christaki, E.; Marques, T.M.; Tosatto, V.; Bianco, G.; Iannaccone, M.; Tsiolakkis, G.; Karagiannis, C.; Maikanti, P.; et al. Severity of RSV infection in Southern European elderly patients during two consecutive winter seasons (2017–2018). *J. Med. Virol.* **2021**, *93*, 5152–5157. [CrossRef] [PubMed]
63. Chorazka, M.; Flury, D.; Herzog, K.; Albrich, W.C.; Vuichard-Gysin, D. Clinical outcomes of adults hospitalized for laboratory confirmed respiratory syncytial virus or influenza virus infection. *PLoS ONE* **2021**, *16*, e0253161. [CrossRef] [PubMed]
64. Celante, H.; Oubaya, N.; Fourati, S.; Beaune, S.; Khellaf, M.; Casalino, E.; Ricard, J.D.; Vieillard-Baron, A.; Heming, N.; Mekontso Dessap, A.; et al. Prognosis of hospitalised adult patients with respiratory syncytial virus infection: A multicentre retrospective cohort study. *Clin. Microbiol. Infect.* **2023**, *29*, e1–e943. [CrossRef] [PubMed]
65. Hämäläinen, A.; Savinainen, E.; Hämäläinen, S.; Sivenius, K.; Kauppinen, J.; Koivula, I.; Patovirta, R.L. Disease burden caused by respiratory syncytial virus compared with influenza among adults: A retrospective cohort study from Eastern Finland in 2017–2018. *BMJ Open* **2022**, *12*, e060805. [CrossRef] [PubMed]
66. Njue, A.; Nuabor, W.; Lyall, M.; Margulis, A.; Mauskopf, J.; Curcio, D.; Kurosky, S.; Gessner, B.D.; Begier, E. Systematic Literature Review of Risk Factors for Poor Outcomes Among Adults with Respiratory Syncytial Virus Infection in High-Income Countries. *Open Forum Infect. Dis.* **2023**, *10*, ofad513. [CrossRef] [PubMed]
67. Schubert, L.; Steininger, J.; Lötsch, F.; Herdina, A.N.; Redlberger-Fritz, M.; Tobudic, S.; Kundi, M.; Strassl, R.; Steininger, C. Surveillance of respiratory syncytial virus infections in adults, Austria, 2017 to 2019. *Sci. Rep.* **2021**, *11*, 8939. [CrossRef] [PubMed]
68. Nolen, L.D.; Seeman, S.; Desnoyers, C.; DeByle, C.; Klejka, J.; Bruden, D.; Rudolph, K.; Gerber, S.I.; Kim, L.; Langley, G.; et al. Respiratory syncytial virus and influenza hospitalizations in Alaska native adults. *J. Clin. Virol.* **2020**, *127*, 104347. [CrossRef] [PubMed]
69. Lee, N.; Chan, M.C.; Lui, G.C.; Li, R.; Wong, R.Y.; Yung, I.M.; Cheung, C.S.; Chan, E.C.; Hui, D.S.; Chan, P.K. High Viral Load and Respiratory Failure in Adults Hospitalized for Respiratory Syncytial Virus Infections. *J. Infect. Dis.* **2015**, *212*, 1237–1240. [CrossRef] [PubMed]
70. Chatzis, O.; Darbre, S.; Pasquier, J.; Meylan, P.; Manuel, O.; Aubert, J.D.; Beck-Popovic, M.; Masouridi-Levrat, S.; Ansari, M.; Kaiser, L.; et al. Burden of severe RSV disease among immunocompromised children and adults: A 10 year retrospective study. *BMC Infect. Dis.* **2018**, *18*, 111. [CrossRef] [PubMed]
71. Holmen, J.E.; Kim, L.; Cikesh, B.; Kirley, P.D.; Chai, S.J.; Bennett, N.M.; Felsen, C.B.; Ryan, P.; Monroe, M.; Anderson, E.J.; et al. Relationship between neighborhood census-tract level socioeconomic status and respiratory syncytial virus-associated hospitalizations in U.S. adults, 2015–2017. *BMC Infect. Dis.* **2021**, *21*, 293. [CrossRef] [PubMed]

72. Pilie, P.; Werbel, W.A.; Riddell, J., 4th; Shu, X.; Schaubel, D.; Gregg, K.S. Adult patients with respiratory syncytial virus infection: Impact of solid organ and hematopoietic stem cell transplantation on outcomes. *Transpl. Infect. Dis.* **2015**, *17*, 551–557. [CrossRef] [PubMed]
73. Azzi, J.M.; Kyvernitakis, A.; Shah, D.P.; El Haddad, L.; Mahajan, S.N.; Ghantoji, S.S.; Heredia-Ariza, E.; Chemaly, R.F. Leukopenia and lack of ribavirin predict poor outcomes in patients with haematological malignancies and respiratory syncytial virus infection. *J. Antimicrob. Chemother.* **2018**, *73*, 3162–3169. [CrossRef] [PubMed]
74. Goldman, C.R.; Sieling, W.D.; Alba, L.R.; Silverio Francisco, R.A.; Vargas, C.Y.; Barrett, A.E.; Phillips, M.; Finelli, L.; Saiman, L. Severe Clinical Outcomes among Adults Hospitalized with Respiratory Syncytial Virus Infections, New York City, 2017–2019. *Public Health Rep.* **2022**, *137*, 929–935. [CrossRef] [PubMed]
75. Lui, G.; Wong, C.K.; Chan, M.; Chong, K.C.; Wong, R.; Chu, I.; Zhang, M.; Li, T.; Hui, D.; Lee, N.; et al. Host inflammatory response is the major marker of severe respiratory syncytial virus infection in older adults. *J. Infect.* **2021**, *83*, 686–692. [CrossRef] [PubMed]
76. Smithgall, M.; Maykowski, P.; Zachariah, P.; Oberhardt, M.; Vargas, C.Y.; Reed, C.; LaRussa, P.; Saiman, L.; Stockwell, M.S. Epidemiology, clinical features, and resource utilization associated with respiratory syncytial virus in the community and hospital. *Influenza Other Respir. Viruses* **2020**, *14*, 247–256. [CrossRef] [PubMed]
77. Widmer, K.; Griffin, M.R.; Zhu, Y.; Williams, J.V.; Talbot, H.K. Respiratory syncytial virus- and human metapneumovirus-associated emergency department and hospital burden in adults. *Influenza Other Respir. Viruses* **2014**, *8*, 347–352. [CrossRef] [PubMed]
78. Duncan, C.B.; Walsh, E.E.; Peterson, D.R.; Lee, F.E.; Falsey, A.R. Risk factors for respiratory failure associated with respiratory syncytial virus infection in adults. *J. Infect. Dis.* **2009**, *200*, 1242–1246. [CrossRef] [PubMed]
79. Widmer, K.; Zhu, Y.; Williams, J.V.; Griffin, M.R.; Edwards, K.M.; Talbot, H.K. Rates of hospitalizations for respiratory syncytial virus, human metapneumovirus, and influenza virus in older adults. *J. Infect. Dis.* **2012**, *206*, 56–62. [CrossRef] [PubMed]
80. Mehta, J.; Walsh, E.E.; Mahadevia, P.J.; Falsey, A.R. Risk factors for respiratory syncytial virus illness among patients with chronic obstructive pulmonary disease. *COPD* **2013**, *10*, 293–299. [CrossRef] [PubMed]
81. Anderson, N.W.; Binnicker, M.J.; Harris, D.M.; Chirila, R.M.; Brumble, L.; Mandrekar, J.; Hata, D.J. Morbidity and mortality among patients with respiratory syncytial virus infection: A 2-year retrospective review. *Diagn. Microbiol. Infect. Dis.* **2016**, *85*, 367–371. [CrossRef] [PubMed]
82. Presser, L.D.; van den Akker, W.M.R.; Meijer, A.; PROMISE investigators. Respiratory Syncytial Virus European Laboratory Network 2022 Survey: Need for Harmonization and Enhanced Molecular Surveillance. *J. Infect. Dis.* **2023**, *229*, S34–S39. [CrossRef] [PubMed]
83. Rozenbaum, M.H.; Begier, E.; Kurosky, S.K.; Whelan, J.; Bem, D.; Pouwels, K.B.; Postma, M.; Bont, L. Incidence of Respiratory Syncytial Virus Infection in Older Adults: Limitations of Current Data. *Infect. Dis. Ther.* **2023**, *12*, 1487–1504. [CrossRef] [PubMed]
84. Carvajal, J.J.; Avellaneda, A.M.; Salazar-Ardiles, C.; Maya, J.E.; Kalergis, A.M.; Lay, M.K. Host Components Contributing to Respiratory Syncytial Virus Pathogenesis. *Front. Immunol.* **2019**, *10*, 2152. [CrossRef] [PubMed]
85. Hall, C.B. Respiratory Syncytial Viruses. In *Mandell, Dougals and Bennet's Principles and Practice of Infectious Diseases*, 7th ed.; Churchill Livingstone: London, UK, 2010; pp. 2207–2227.
86. Colosia, A.; Costello, J.; McQuarrie, K.; Kato, K.; Bertzos, K. Systematic literature review of the signs and symptoms of respiratory syncytial virus. *Influenza Other Respir. Viruses* **2023**, *17*, e13100. [CrossRef] [PubMed]
87. Riccò, M.; Corrado, S.; Palmieri, S.; Marchesi, F. Respiratory Syncytial Virus: A Systematic Review and Meta-Analysis of Tomographic Findings (2000–2022). *Children* **2023**, *10*, 1169. [CrossRef] [PubMed]
88. Shah, D.P.; Ghantoji, S.S.; Shah, J.N.; El Taoum, K.K.; Jiang, Y.; Popat, U.; Hosing, C.; Rondon, G.; Tarrand, J.J.; Champlin, R.E.; et al. Impact of aerosolized ribavirin on mortality in 280 allogeneic haematopoietic stem cell transplant recipients with respiratory syncytial virus infections. *J. Antimicrob. Chemother.* **2013**, *68*, 1872–1880. [CrossRef] [PubMed]
89. Trang, T.P.; Whalen, M.; Hilts-Horeczko, A.; Doernberg, S.B.; Liu, C. Comparative effectiveness of aerosolized versus oral ribavirin for the treatment of respiratory syncytial virus infections: A single-center retrospective cohort study and review of the literature. *Transpl. Infect. Dis.* **2018**, *20*, e12844. [CrossRef] [PubMed]
90. Lee, F.E.; Walsh, E.E.; Falsey, A.R. The effect of steroid use in hospitalized adults with respiratory syncytial virus-related illness. *Chest* **2011**, *140*, 1155–1161. [CrossRef] [PubMed]
91. Jeannoël, M.; Lina, G.; Rasigade, J.P.; Lina, B.; Morfin, F.; Casalegno, J.S. Microorganisms associated with respiratory syncytial virus pneumonia in the adult population. *Eur. J. Clin. Microbiol. Infect. Dis.* **2019**, *38*, 157–160. [CrossRef] [PubMed]
92. Zhang, Y.; Zhao, J.; Zou, X.; Fan, Y.; Xiong, Z.; Li, B.; Wang, C.; Li, H.; Han, J.; Liu, X.; et al. Severity of influenza virus and respiratory syncytial virus coinfections in hospitalized adult patients. *J. Clin. Virol.* **2020**, *133*, 104685. [CrossRef] [PubMed]
93. Godefroy, R.; Giraud-Gatineau, A.; Jimeno, M.T.; Edouard, S.; Meddeb, L.; Zandotti, C.; Chaudet, H.; Colson, P.; Raoult, D.; Cassir, N. Respiratory Syncytial Virus Infection: Its Propensity for Bacterial Coinfection and Related Mortality in Elderly Adults. *Open Forum Infect. Dis.* **2020**, *7*, ofaa546. [CrossRef] [PubMed]
94. Park, S.Y.; Kim, T.; Jang, Y.R.; Kim, M.C.; Chong, Y.P.; Lee, S.O.; Choi, S.H.; Kim, Y.S.; Woo, J.H.; Kim, S.H. Factors predicting life-threatening infections with respiratory syncytial virus in adult patients. *Infect. Dis.* **2017**, *49*, 333–340. [CrossRef] [PubMed]

95. Osei-Yeboah, R.; Johannesen, C.K.; Egeskov-Cavling, A.M.; Chen, J.; Lehtonen, T.; Fornes, A.U.; Paget, J.; Fischer, T.K.; Wang, X.; Nair, H.; et al. Respiratory syncytial virus-associated hospitalisation in adults with comorbidities in two European countries: A modelling study. *J. Infect. Dis.* **2023**, *229*, S70–S77. [CrossRef] [PubMed]
96. Almeida, A.; Boattini, M.; Christaki, E.; Moreira Marques, T.; Moreira, I.; Cruz, L.; Tosatto, V.; Antão, D.; Bianco, G.; Iannaccone, M.; et al. Comparative virulence of seasonal viruses responsible for lower respiratory tract infections: A southern European multi-centre cohort study of hospital admissions. *Infection* **2021**, *49*, 483–490. [CrossRef] [PubMed]
97. Boattini, M.; Almeida, A.; Christaki, E.; Cruz, L.; Antão, D.; Moreira, M.I.; Bianco, G.; Iannaccone, M.; Tsiolakkis, G.; Khattab, E.; et al. Influenza and respiratory syncytial virus infections in the oldest-old continent. *Eur. J. Clin. Microbiol. Infect. Dis.* **2020**, *39*, 2085–2090. [CrossRef] [PubMed]
98. Casiano-Colón, A.E.; Hulbert, B.B.; Mayer, T.K.; Walsh, E.E.; Falsey, A.R. Lack of sensitivity of rapid antigen tests for the diagnosis of respiratory syncytial virus infection in adults. *J. Clin. Virol.* **2003**, *28*, 169–174. [CrossRef] [PubMed]
99. Chartrand, C.; Tremblay, N.; Renaud, C.; Papenburg, J. Diagnostic Accuracy of Rapid Antigen Detection Tests for Respiratory Syncytial Virus Infection: Systematic Review and Meta-analysis. *J. Clin. Microbiol.* **2015**, *53*, 3738–3749. [CrossRef] [PubMed]
100. Onwuchekwa, C.; Moreo, L.M.; Menon, S.; Machado, B.; Curcio, D.; Kalina, W.; Atwell, J.E.; Gessner, B.D.; Siapka, M.; Agarwal, N.; et al. Underascertainment of Respiratory Syncytial Virus Infection in Adults Due to Diagnostic Testing Limitations: A Systematic Literature Review and Meta-analysis. *J. Infect. Dis.* **2023**, *228*, 173–184. [CrossRef] [PubMed]
101. Available online: https://www.who.int/teams/global-influenza-programme/global-respiratory-syncytial-virus-surveillance (accessed on 20 December 2023).
102. Available online: https://www.who.int/teams/global-influenza-programme/global-respiratory-syncytial-virus-surveillance/case-definitions (accessed on 20 December 2023).
103. Available online: https://www.who.int/teams/global-influenza-programme/surveillance-and-monitoring/case-definitions-for-ili-and-sari (accessed on 20 December 2023).
104. Available online: https://www.who.int/teams/global-influenza-programme/global-respiratory-syncytial-virus-surveillance/sampling-strategy-for-rsv-testing (accessed on 20 December 2023).
105. Available online: https://www.who.int/teams/global-influenza-programme/global-respiratory-syncytial-virus-surveillance/collection-transport-and-storage (accessed on 20 December 2023).
106. Osei-Yeboah, R.; Spreeuwenberg, P.; Del Riccio, M.; Fischer, T.K.; Egeskov-Cavling, A.M.; Bøås, H.; van Boven, M.; Wang, X.; Lehtonen, T.; Bangert, M.; et al. Estimation of the Number of Respiratory Syncytial Virus-Associated Hospitalizations in Adults in the European Union. *J. Infect. Dis.* **2023**, *228*, 1539–1548. [CrossRef] [PubMed]
107. Rozenbaum, M.H.; Judy, J.; Tran, D.; Yacisin, K.; Kurosky, S.K.; Begier, E. Low Levels of RSV Testing Among Adults Hospitalized for Lower Respiratory Tract Infection in the United States. *Infect. Dis. Ther.* **2023**, *12*, 677–685. [CrossRef] [PubMed]
108. Korsten, K.; Adriaenssens, N.; Coenen, S.; Butler, C.C.; Verheij, T.J.M.; Bont, L.J.; Wildenbeest, J.G.; RESCEU Investigators. World Health Organization Influenza-Like Illness Underestimates the Burden of Respiratory Syncytial Virus Infection in Community-Dwelling Older Adults. *J. Infect. Dis.* **2022**, *226*, S71–S78. [CrossRef] [PubMed]
109. Pebody, R.; Moyes, J.; Hirve, S.; Campbell, H.; Jackson, S.; Moen, A.; Nair, H.; Simões, E.A.F.; Smith, P.G.; Wairagkar, N.; et al. Approaches to use the WHO respiratory syncytial virus surveillance platform to estimate disease burden. *Influenza Other Respir. Viruses* **2020**, *14*, 615–621. [CrossRef] [PubMed]
110. Cawcutt, K.; Kalil, A.C. Pneumonia with bacterial and viral coinfection. *Curr. Opin. Crit. Care* **2017**, *23*, 385–390. [CrossRef] [PubMed]
111. Pacheco, G.A.; Gálvez, N.M.S.; Soto, J.A.; Andrade, C.A.; Kalergis, A.M. Bacterial and Viral Coinfections with the Human Respiratory Syncytial Virus. *Microorganisms* **2021**, *9*, 1293. [CrossRef] [PubMed]
112. Mazur, N.I.; Terstappen, J.; Baral, R.; Bardají, A.; Beutels, P.; Buchholz, U.J.; Cohen, C.; Crowe, J.E., Jr.; Cutland, C.L.; Eckert, L.; et al. Respiratory syncytial virus prevention within reach: The vaccine and monoclonal antibody landscape. *Lancet Infect. Dis.* **2023**, *23*, e2–e21. [CrossRef] [PubMed]
113. Available online: https://www.path.org/our-impact/resources/rsv-vaccine-and-mab-snapshot/ (accessed on 29 January 2024).
114. Papi, A.; Ison, M.G.; Langley, J.M.; Lee, D.G.; Leroux-Roels, I.; Martinon-Torres, F.; Schwarz, T.F.; van Zyl-Smit, R.N.; Campora, L.; Dezutter, N.; et al. Respiratory Syncytial Virus Prefusion F Protein Vaccine in Older Adults. *N. Engl. J. Med.* **2023**, *388*, 595–608. [CrossRef] [PubMed]
115. Walsh, E.E.; Pérez Marc, G.; Zareba, A.M.; Falsey, A.R.; Jiang, Q.; Patton, M.; Polack, F.P.; Llapur, C.; Doreski, P.A.; Ilangovan, K.; et al. Efficacy and Safety of a Bivalent RSV Prefusion F Vaccine in Older Adults. *N. Engl. J. Med.* **2023**, *388*, 1465–1477. [CrossRef] [PubMed]
116. Walsh, E.E.; Falsey, A.R.; Scott, D.A.; Gurtman, A.; Zareba, A.M.; Jansen, K.U.; Gruber, W.C.; Dormitzer, P.R.; Swanson, K.A.; Radley, D.; et al. A Randomized Phase 1/2 Study of a Respiratory Syncytial Virus Prefusion F Vaccine. *J. Infect. Dis.* **2022**, *225*, 1357–1366. [CrossRef] [PubMed]
117. Wilson, E.; Goswami, J.; Baqui, A.H.; Doreski, P.A.; Perez-Marc, G.; Zaman, K.; Monroy, J.; Duncan, C.J.A.; Ujiie, M.; Rämet, M.; et al. Efficacy and Safety of an mRNA-Based RSV PreF Vaccine in Older Adults. *N. Engl. J. Med.* **2023**, *389*, 2233–2244. [CrossRef]
118. Available online: https://clinicaltrials.gov/study/NCT05585632?cond=mRNA-1230&rank=1#study-overview (accessed on 29 January 2024).

119. Cockerill, G.S. JNJ-5371678, Defining a Role for Fusion Inhibitors in the Treatment of Respiratory Syncytial Virus. *J. Med. Chem.* **2020**, *63*, 8043–8045. [CrossRef] [PubMed]
120. Ahmad, A.; Eze, K.; Noulin, N.; Horvathova, V.; Murray, B.; Baillet, M.; Grey, L.; Mori, J.; Adda, N. EDP-938, a Respiratory Syncytial Virus Inhibitor, in a Human Virus Challenge. *N. Engl. J. Med.* **2022**, *386*, 655–666. [CrossRef] [PubMed]
121. Kim, H.W.; Canchola, J.G.; Brandt, C.D.; Pyles, G.; Chanock, R.M.; Jensen, K.; Parrott, R.H. Respiratory syncytial virus disease in infants despite prior administration of antigenic inactivated vaccine. *Am. J. Epidemiol.* **1969**, *89*, 422–434. [CrossRef] [PubMed]
122. Melgar, M.; Britton, A.; Roper, L.E.; Talbot, H.K.; Long, S.S.; Kotton, C.N.; Havers, F.P. Use of Respiratory Syncytial Virus Vaccines in Older Adults: Recommendations of the Advisory Committee on Immunization Practices—United States, 2023. *Morb. Mortal. Wkly. Rep. (MMWR)* **2023**, *72*, 793–801. [CrossRef] [PubMed]

viruses

Article

Severe Pneumonia Caused by Respiratory Syncytial Virus and Adenovirus in Children from 2 to 24 Months at Children's Hospital 1 in Ho Chi Minh City, Vietnam

Suong Thi Thu Nguyen [1,2,*], Tuan Anh Tran [1,3] and Giau Van Vo [1,2,4,*]

[1] School of Medicine, Vietnam National University—Ho Chi Minh City (VNU-HCM), Ho Chi Minh City 710000, Vietnam
[2] Vietnam National University—Ho Chi Minh City (VNU-HCM), Ho Chi Minh City 700000, Vietnam
[3] Children's Hospital 1, Ho Chi Minh City 710000, Vietnam
[4] Research Center for Genetics and Reproductive Health (CGRH), School of Medicine, Vietnam National University, Ho Chi Minh City (VNU-HCM), Ho Chi Minh City 70000, Vietnam
* Correspondence: nttsuong@medvnu.edu.vn (S.T.T.N.); vvgiau@medvnu.edu.vn (G.V.V.)

Abstract: In Vietnam, due to the lack of facilities to detect respiratory viruses from patients' specimens, there are only a few studies on the detection of viral pathogens causing pneumonia in children, especially respiratory syncytial virus (RSV) and adenovirus (Adv). Here, we performed a cross-sectional descriptive prospective study on 138 children patients from 2 to 24 months old diagnosed with severe pneumonia hospitalized at the Respiratory Department of Children's Hospital 1 from November 2021 to August 2022. The number of patients selected in this study was based on the formula n = ([Z(1 − α/2)]2 × P [1 − P])/d2, with $\alpha = 0.05$, $p = 0.5$, and d = 9%, and the sampling technique was convenient sampling until the sample size was met. A rapid test was used to detect RSV and Adv from the nasopharyngeal swabs and was conducted immediately after the patient's hospitalization. Laboratory tests were performed, medical history interviews were conducted, and nasotracheal aspirates were collected for multiplex real-time PCR (MPL-rPCR) to detect viral and bacterial pathogens. The results of the rapid test and the MPL-rPCR in the detection of both pathogens were the same at 31.9% (44/138) for RSV and 8.7% (7/138) for Adv, respectively. Using MPL-rPCR, the detection rate was 21% (29/138) for bacterial pathogens, 68.8% (95/138) for bacterial–viral co-infections, and 6.5% (9/138) for viral pathogens. The results showed few distinctive traits between RSV-associated and Adv-associated groups, and the Adv group children were more prone to bacterial infection than those in the RSV group. In addition, the Adv group experienced a longer duration of treatment and a higher frequency of re-hospitalizations compared to the RSV group. A total of 100% of Adv infections were co-infected with bacteria, while 81.82% of RSV co-infected with bacterial pathogens ($p = 0.000009$). This study might be one of the few conducted in Vietnam aimed at identifying viral pathogens causing severe pneumonia in children.

Keywords: respiratory syncytial virus (RSV); adenovirus (Adv); severe pneumonia; multiplex real-time PCR; Nasal Continuous Positive Airway Pressure (NCPAP)

Citation: Nguyen, S.T.T.; Tran, T.A.; Vo, G.V. Severe Pneumonia Caused by Respiratory Syncytial Virus and Adenovirus in Children from 2 to 24 Months at Children's Hospital 1 in Ho Chi Minh City, Vietnam. *Viruses* **2024**, *16*, 410. https://doi.org/10.3390/v16030410

Academic Editors: Emanuele Castagno and Irene Raffaldi

Received: 29 January 2024
Revised: 2 March 2024
Accepted: 4 March 2024
Published: 7 March 2024

1. Introduction

Community-acquired pneumonia is a common respiratory disease affecting all ages, especially children under 5 years old. The incidence and hospitalization rate due to pneumonia remains at a high level, given that it is the leading cause of death in children, especially in low-income countries like Cambodia, Laos, Vietnam, China, and the Philippines [1]. The etiology of pneumonia encompasses various infectious organisms. However, viruses account for a significant proportion, as shown by the study by Kouni et al. [2], in which the co-infection of viruses in children with acute respiratory infection accounted for 42.5%. According to numerous studies and the literature, viruses were the most common cause of acute lower respiratory tract infections. Notably, respiratory syncytial virus (RSV) was particularly prevalent in infants, especially those under 6 months of age. In addition, other viruses such as adenovirus (Adv), influenza, parainfluenza, rhinovirus, and human metapneumovirus were also important causes, with Adv being the second most common after RSV [2,3]. Rodríguez-Martínez et al. from 2009 to 2011 documented 2.267 children with lower respiratory infections [3], of which 87,8% were identified as being caused by RSV, while 9% were attributed to Adv infection; 3.1% were found to be the result of both RSV and Adv infection. In some cases, Adv infections resulted in a more severe illness than RSV, leading to life-threatening and prolonged pneumonia and necessitating admission to the intensive care unit for respiratory support [4]. In addition, Adv is highly contagious and there is no preventive vaccine available to mitigate its spread at present. Hence, we carried out this study with the primary aim of assessing the occurrence of viral and bacterial–viral co-infection in children aged 2 months to 24 months of age with severe community-acquired pneumonia requiring oxygen therapy in the emergency room of the respiratory department of Children's Hospital 1. Additionally, this study sought to examine the characteristics of clinical, subclinical, and microbiological characteristics along with treatment outcomes of severe viral, bacterial, and bacterial–viral co-infection pneumonia in this population.

2. Methods

2.1. Study Design and Population

This is a cross-sectional descriptive prospective study on 138 children patients from 2 to 24 months old diagnosed with severe pneumonia hospitalized at the Respiratory Department of Children's Hospital 1 from November 2021 to August 2022. The inclusion criteria were (i) age from 2 to 24 months old; (ii) clinical signs including cough, shortness of breath, tachypnea according to age, chest indrawing, and indications for oxygen (WHO 2016); and (iii) parenchymal injury detected by a chest radiograph. On the other hand, following the guidelines for the diagnosis and management of bronchiolitis by the WHO in 2016, we excluded children who had respiratory failure due to being diagnosed with bronchiolitis. Some specific characteristic features include a viral upper respiratory prodrome followed by increased respiratory effort (e.g., tachypnea, nasal, flaring, chest retractions) and wheezing and/or crackles in children younger than two years of age, as well as chest radiographs that are not damaging to lung parenchyma. The pediatric patients infected with COVID-19 confirmed by a rapid test and/or PCR, or patients whose nasal tracheal aspirate (NTA) could not be collected or whose NTA could not meet the Bartlett score [5] to indicate the high quality of the expectorated sputum based on the Gram stain smear of the specimen at the laboratory right after receiving the sample, or patients who had received intravenous antibiotics within 24 h before admission were excluded from this study. The total number of the patients selected in this study was based on the formula $n = ([Z_{(1-\alpha/2)}]^2 \times P\,[1-P])/d^2$, with $\alpha = 0.05$, $p = 0.5$, and d = 9%; then, $n = 118$. The sampling technique used in this study was convenient sampling until the sample size was met.

All pediatric patients who met the mentioned criteria were interviewed for medical history and clinical examination. Laboratory tests, including complete blood count, CRP, chest X-ray, and nasopharyngeal swab to quickly detect respiratory syncytial virus (RSV) and adenovirus (Adv), were carried out. The NTA samples were collected within 24 h of admission. The chest radiograph was reviewed by the head physician of the radiology department. Disease progression, laboratory results, and antibiotic therapy were recorded. To analyze NTA samples, 1 mL of each sample collected from patients was sent to the laboratory for multiplex real-time PCR (MPL rPCR) testing to detect the microorganism pathogens with the protocol and materials, as in a previous study [6,7]. In addition, the collected NTA samples were also tested for RSV and Adv antigens using the "FUJI DRI-CHEM IMMUNO AG RSV/Adv" system based on the immuneoelectrophoresis principle (95% sensitivity and 100% specificity compared to standard PCR). Patients with RSV and Adv pneumonia were characterized in Table 1.

Table 1. Some clinical characteristics of RSV and Adv among children with pneumonia.

Clinical Features		RSV (n = 44)	Adv (n = 7)	p-Value
Temperature	<38 °C	35 (79.5%)	2 (28.5%)	0.001
	38 °C–38.5 °C	0	1 (14.3%)	
	38.6 °C–39 °C	9 (20.5%)	3 (42.9%)	
	>39 °C	0	1 (14.3%)	
SpO_2	<85%	1 (2.3%)	1 (14.3%)	0.518
	85%–90%	22 (50%)	3 (42.9%)	
	91%–94%	21 (47.4%)	3 (42.9%)	
Chest indrawing	Indrawing	36 (81.8%)	2 (28.6%)	0.001
	Severe indrawing	8 (18.2)	5 (71.9%)	
Wheezing		34 (77.3%)	5 (71.4%)	0.042
Wheezes—rhonchi		34 (77.3%)	6 (85.7%)	0.008

2.2. Statistical Analysis

The collected data were processed using SPSS 20.0 software for subsequent statistical analysis. Categorical variables were compared using χ^2 or Fisher's exact test. The statistical significance was defined as $p < 0.05$. Data are expressed as the number of cases and the percentage.

2.3. Ethics Statement

This study was approved by the Ethics Committee of Children's Hospital 1 in Ho Chi Minh City on 7 December 2022 (No. 269/GCN-BVND1). Stringent measures were in place to ensure patient privacy and data confidentiality. Personal identifying information was anonymized during data analysis, and all research procedures adhered to the ethical guidelines and regulations applicable in Vietnam. The medical record information of the patients participating in this study was used for research purposes only and not for any other purpose.

3. Results

From November 2021 to August 2022, there were 138 pediatric patients from 2 to 24 months old admitted to the Respiratory Department who met the inclusion criteria. Among these patients, 93 were boys and 33 were girls; the mean age was 7.39 months old with the youngest being 2 months old and the oldest being 23 months old. The most common age range was from 2 to 6 months old (59.4%), followed by 6 to 12 months old (21%) and over 12 months old (18.8%). There were 32 cases (23.2%) of low birth weight or preterm births and 29 cases (21%) that had postpartum respiratory failure.

All of these patients underwent rapid tests for RSV and Adv. The results from the rapid test showed seven cases positive for Adv, forty-four cases positive with RSV, and eighty-seven cases negative with both Adv and RSV. The NTA for multiplex real-time PCR testing was taken within the first 24 h of admission and sent quickly to the laboratory. The results of the MPL rPCR in the detection of Adv and RSV were completely consistent with those of the rapid test. The MPL rPCR results also showed that there were twenty-nine cases positive for bacterial pathogens (21%), ninety-five cases positive for bacterial and viral pathogens (68.8%), nine cases positive for viral pathogens (6.5%), and five cases negative for both bacterial and viral pathogens (3.7%). Table 2 shows the results of the MPL rPCR in the detection of bacterial and viral pathogens in 138 NTA samples.

Table 2. The bacterial and viral pathogen detection rate by MPL rPCR.

Bacterial Pathogen	n	%	Viral Pathogen	n	%
Streptococcus pneumoniae	69	49.8	Respiratory syncytial virus (RSV)	44	31.9
MRSA	18	13.1	Adenovirus (Adv)	7	5.1
MRSE	15	10.6	Rhinovirus	12	8.7
Staphylococcus epidermidis	6	4.1	Bocavirus	28	20.3
Hemophilus influenza UT	12	9	Influenza virus A	1	0.7
Moraxella catarrhalis	7	4.9	Parainfluenza virus type 3	21	15.2
Mycoplasma pneumoniae	8	5.7	Cytomegalovirus (CMV)	35	25.4
Chlamydia trachomatis	18	13.1	Epstein–Barr virus (EBV)	5	3.6
Burkholderia cepacia	5	3.3	% is the percentage of pathogen detection in 138 NTA samples collected from patients		
Escherichia coli	26	18.8	MRSA: methicillin-resistant *Staphylococcus aureus*		
Klebsiella pneumoniae	16	11.3	MRSE: methicillin-resistant *Staphylococcus epidermidis*		
Acinetobacter baumannii	9	6.5			
Pseudomonas aeruginosa	9	6.5			

Table 2 data indicate that among the bacterial pathogens, *S. pneumoniae* was the pathogen with the highest detection ratio (49.8%). Among the atypical bacterial pathogens, *Chlamydia trachomatis* was detected with a ratio of 13.1%, higher than that of *Mycoplasma pneumoniae* (5.7%). RSV exhibited the highest detection ratio (31.9%) among the viral pathogens. CMV had a high detection ratio of 25.4%, although this viral pathogen is often regarded as the etiological pathogen of pneumonia in the immunocompromised host; CMV is also considered a potential pathogen of severe pneumonia in children in non-HIV-infected children [8].

In order to analyze the differences between severe pneumonia associated with RSV and Adv, the related laboratory and treatment characteristics were also reported and shown in Tables 3 and 4.

Table 3. Laboratory characteristics of severe pneumonia associated with RSV and Adv.

Laboratory Features		RSV (n = 44)	Adv (n = 7)	p-Value
WBC	≤15,000 cell/mm^3	32 (72.7%)	2 (28.6%)	0.047
	>15,000 cell/mm^3	12 (27.3%)	5 (71.4%)	
Neutrophil count	<8000 cell/ mm^3	30 (68.2)	2 (28.6%)	0.064
	≥8000 cell/mm^3	14 (31.8%)	5 (71.4%)	
CRP	≤35 mg/L	34 (77.3%)	2 (28.6%)	0.012
	>35 mg/L	10 (22.7%)	5 (71.4%)	
Elevated liver enzyme *	Yes	3 (6.8%)	4 (57.1%)	0.00001 **
	No	41(93.2%)	3(42.9%)	
Chest X-ray	Bilateral infiltrates	18 (40.9%)	6 (85.7%)	0.359 **
	One-sided infiltrates	14 (31.8%)	0	
	Consolidation	3 (6.9%)	0	
	Right upper lobe collapse	8 (18.2%)	1 (14.3%)	
	Pneumonia–pleural effusion	1 (2.3%)	0	

Elevated liver enzyme *: ALT > 45U/L and/or AST > 60U/L, **: Fisher.

Table 4. Treatment characteristics of severe pneumonia associated with RSV and Adv.

Treatment Characteristic		RSV (n = 44)	Adv (n = 7)	p-Value
Duration *	≤7 days	4 (9.1%)	0	0.168
	8–14 days	26 (59.1%)	1 (14.3%)	
	15–30 days	10 (22.7%)	3 (42.9%)	
	>30 days	4 (9.1%)	3 (42.9%)	
Antibiotic change	1 time	20 (45.5%)	3 (42.9%)	0.18
	2 times	4 (9.1%)	2 (28.6%)	
	≥3 times	2 (4.5%)	1 (14.3%)	
Respiratory support	Oxy/cannula	20 (45.5%)	2(28.6%)	0.619
	HFNC *	1 (2.3%)	0	
	NCPAP	20 (45.5%)	4 (57.1%)	
	ETT ventilation	3 (6.8%)	1 (14.3%)	
Re-hospitalization **	1 time	8 (18.2%)	3 (42.9%)	0.002
	2 times	0	0	
	≥3 times	0	2 (28.6%)	

HFNC: high-flow nasal cannula, *NCPAP*: Nasal Continuous Positive Airway Pressure, ETT ventilation: endotracheal tube ventilation. * Duration is the total number of days that the patient was treated in the hospital. ** Re-hospitalization is hospitalization again because of recurrent pneumonia within 1 month after discharge from the hospital.

In Table 3, the obtained data show that (i) a white blood cell (WBC) count of more than 15,000 cell/mm^3 was mainly seen in severe pneumonia associated with Adv; however, in the RSV group, WBC count was ≤15,000 cell/mm^3, and the difference was statistically significant with p = 0.047 and χ^2 = 6.003. (ii) Severe pneumonia cases with C-reactive protein (CRP) more than 35 mg/L were encountered mainly in the Adv group, while CRP $\leq$ 35 mg/L was mainly in the RSV group, and the difference was statistically significant with p = 0.012 and χ^2 = 8.795. (iii) Elevated liver enzymes were mainly observed in the Adv group (57.1%).

The obtained data in Table 4 show that (i) most patients in the RSV group were treated in less than 2 weeks, while most of the Adv patients were hospitalized for over 30 days, and the difference was statistically significant with p = 0.002 and χ^2 = 18,375. (ii) Some patients in both groups were re-hospitalized once; however, readmission more than three times was only recorded in the Adv group, including one case that progressed to PIBO (postinfectious bronchiolitis obliterans), and the difference was statistically significant with p = 0.0021 and χ^2 = 37,059.

Why did the two groups of children with severe pneumonia exhibit differences in laboratory findings and treatment outcomes in which the Adv group was more prone to bacterial infection (high white blood cell count, high CRP, increased liver enzymes), required a longer duration of treatment, and experienced more frequent re-hospitalizations than the RSV group? To answer this question, it was necessary to analyze the bacterial co-infection of the two groups. The analyzed results showed that in the RSV group, there were 36 cases of bacterial pathogen co-infection, accounting for 81.82% (36/44), while in the Adv group, 100% (7/7) were co-infected with the bacterial pathogens. Although the number of cases in the Adv group in this study was relatively low (only seven cases), with the statistical analysis using the binomial test, this difference was statistically significant with p = 0.000009.

4. Discussions

The present study shows that RSV and adenoviruses are significant causes of acute severe pneumonia in infants and young children in low- and middle-income countries, including Vietnam. Additionally, the findings of the present study suggest that the Adv group children were more prone to bacterial infection than those in the RSV group. Also, patients co-infected with other pathogenic bacteria were more frequently observed in both groups. Of all the viral infections, RSV was the most often detected, affecting 31.9% of patients; rhinovirus was detected in only 8.7% of cases. According to a prior Vietnamese study, 632 infants under the age of two who had community-acquired pneumonia had up to 48% of RSV and 6% of Adv [9]. Compared to our study, which did not include the rainy season, this one may have included two RSV infection episodes (the rainy season) from May 2009 to December 2010. The RSV detection ratio in our study was greater than that of a study conducted on 1082 hospitalized children with lower respiratory tract infections between April 2010 and May 2011, which revealed a 23.8% RSV detection rate [10]. Benjamin M. Althouse et al. observed 15.2% (455/2998) plus influenza A virus 6.1% and rhinovirus 19% in a distinct study conducted between 2007 and 2012 on children hospitalized for acute respiratory virus infection in the city of Nha Trang, located in central Vietnam. Nevertheless, the detection ratio of Adv in this investigation was a mere 2.9% [11]. Naturally, because of the high humidity during the rainy season, the highest rate of infectious respiratory infections was seen. Given that Nha Trang, which is directly overlooking the sea, has a more temperate dry climate with a shorter rainy season than Ho Chi Minh City, which is located further inland and experiences intense heat and humidity during the rainy season, the differences are most likely due to the different geographic areas with different climates.

In our study, the rates of severe RSV-associated and advanced pneumonia groups requiring mechanical ventilation (NTT, CPAP, oxygen/cannula, etc.) were comparable. On the other hand, the Adv group saw a greater rate of re-hospitalization than the RSV group. These results are consistent with a number of evaluations of the literature that indicated that RSV was the most common virus responsible for lower respiratory tract infections in babies, with Adv coming in second [4,11–14]. These studies also revealed that 10% of hospitalized RSV-infected newborns may go on to acquire asthma later in life, and that wheezing from lower respiratory tract infections caused by RSV infection could last for a long time [4,11–14]. However, following a 5-year follow-up period, 50% of individuals with severe Adv-associated pneumonia resulted in the development of PIBO (postinfectious bronchiolitis obliterans) [4,11–14].

Numerous studies revealed that between 50 and 90% of lower respiratory tract infection cases in children under the age of five were caused by viral pathogens. The majority of viral–bacterial co-infected pneumonia cases affected children younger than 2 years old [2,13,14]. These data were entirely consistent with our study's findings, which showed that 94 out of 138 cases (68.1%) had bacterial–viral co-infection and 75.4% (104/138) had viral pathogens detected. Children who obtain community-acquired pneumonia frequently have both bacterial and viral co-infection, which exacerbates the illness and raises the death risk. The results of our investigation were fully consistent with the incidence of bacterial and viral co-infection in pneumonia, which can reach up to 68% of hospitalized patients [2,11]. Numerous investigations also revealed that *S. aureus* and *S. pneumoniae* were the most frequently found bacterial pathogens co-infected with viral pathogens [9,12,13]. This finding was consistent with our investigation, as the two bacterial pathogens detected with the highest ratio were *S. pneumoniae* and *staphylococci*.

In the current study, the majority of patients with Adv infection had SpO2 levels below 85% at admission, followed by SpO2 levels between 85% and 90%. Additionally, 71.4% of patients had increased work of breathing with severe chest withdrawal, and 28.6% had chest withdrawal. On the other hand, a significant number of patients in the group infected with RSV had an initial SpO2 level between 85% and 90% (50%), 47.7% had a level between 91% and 94%, and only 2.3% of cases had a level below 85%. When it came to the level of the withdrawal chest, most patients in the Adv-infected group (71.9%) had severe indrawing, but many patients in the RSV group only experienced less severe indrawing chests. Zampoli's study [15] showed that hypoxemia affected 70.9% of children with pneumonia related to Adv, whereas Li Min Lim's study [14] showed that respiratory failure affected 67.2% of the children. As a result, when severe Adv-infected pneumonia vs. RSV worsens, it should be recognized and treated right away [3,12,15].

While several of the children in the Adv-infected group of our study received treatment for longer than 30 days, most of the RSV-infected children were hospitalized for less than two weeks. While most patients in the RSV group changed antibiotics just once throughout therapy, the majority of the Adv group required antibiotic changes three times (14.3%). As previously mentioned, the RSV group mostly needed oxygen or cannulas for respiratory assistance, but the Adv-infected group needed NCPAP or other cutting-edge techniques. Children with advanced disease were susceptible to not responding to the first round of antibiotics (cefotaxime or ceftriaxone), which would mean a lengthier course of treatment. In our investigation, every instance with advanced pneumonia had a co-infection with at least one species of bacteria, necessitating the use of antibiotics. In addition, our patients in both groups were readmitted to the hospital; however, the Adv group was the only one to be readmitted three times, with one case progressing to pneumonia after contracting an infection (based on clinical and chest CT scan data, ruling out other variables that could induce interstitial lung damage). A 5-year follow-up study was carried out by Rodriguez [3] on 38 children who were hospitalized in 1998 during an outbreak of Adv pneumonia in Santiago, Chile. According to the study, almost 50% of the cases went on to develop PIBO, and the requirements for oxygen therapy, mechanical ventilation, and ICU (intensive care unit) hospitalization were linked to this complication. While the incidence of pneumonia due to Adv was only 25–12%, serotypes 3, 7, and 14 have the potential to cause fatal necrotizing pneumonia [3,15–23]. This could be one of the factors contributing to severe pneumonia linked to Adv in pediatric patients. The small patient population and the fact that this study was carried out in a specialist hospital are its two primary limitations. Additionally, due to the small group sizes of each respiratory virus, we were unable to ascertain the independent effect of each respiratory virus concurrently diagnosed with the illnesses. Lastly, it was not possible to evaluate the effect of co-infection on the long-term effects of RSV and Adv infection.

5. Conclusions

Bacterial–viral co-infection pneumonia has a high prevalence in children under 2 years of age, with *S. pneumoniae* infection and RSV being the primary contributors. Severe CAP with wheezing is suggestive of RSV infection, while severe CAP requiring oxygen with a high-grade and prolonged fever, elevated WBC, elevated liver enzymes, and prolonged treatment time suggests Adv in conjunction with drug-resistant bacteria. To aid clinical physicians in stratifying management plans and to help limit the spread of these two viruses, rapid tests and MPL-rPCR to detect RSV and Adv should be indicated in patients with severe CAP requiring oxygen.

Author Contributions: S.T.T.N. performed the experiments, analyzed the data, and wrote the first draft of the manuscript. S.T.T.N. and T.A.T. collected and analyzed the patient's clinical data and managed the patient's care. S.T.T.N., T.A.T. and G.V.V. conceived this study, analyzed the data, and finalized the manuscript. S.T.T.N. and G.V.V. revised the manuscript. All authors have read and agreed to the published version of the manuscript.

Funding: This research was funded by the Vietnam National University HoChiMinh City (VNU-HCM) under grant number C2022-44-02.

Institutional Review Board Statement: Ethic Committee Name: Children's Hospital 1 in Ho Chi Minh City; approval code No. 269/GCN-BVND1; approval date 7 December 2022.

Informed Consent Statement: Written informed consent was obtained from all patients participating in this study.

Data Availability Statement: The data that support the findings of this study are available from the corresponding author upon reasonable request.

Acknowledgments: We would like to acknowledge all participants who enrolled in this study and their parents and families. We would like to acknowledge our supporting colleagues at the School of Medicine, Vietnam National University Ho Chi Minh City (VNU-HCM), and Children's Hospital 1.

Conflicts of Interest: The authors declare that the research was conducted in the absence of any commercial or financial relationships that could be construed as potential conflicts of interest.

References

1. Nguyen, T.K.P.; Tran, T.H.; Roberts, C.L.; Fox, G.J.; Graham, S.M.; Marais, B.J. Risk factors for child pneumonia—Focus on the Western Pacific Region. *Paediatr. Respir. Rev.* **2017**, *21*, 95–101. [CrossRef]
2. Kouni, S.; Karakitsos, P.; Chranioti, A.; Theodoridou, M.; Chrousos, G.; Michos, A. Evaluation of viral co-infections in hospitalized and non-hospitalized children with respiratory infections using microarrays. *Clin. Microbiol. Infect.* **2013**, *19*, 772–777. [CrossRef]
3. Rodríguez-Martínez, C.E.; Rodríguez, D.A.; Nino, G. Respiratory syncytial virus, adenoviruses, and mixed acute lower respiratory infections in children in a developing country. *J. Med. Virol.* **2015**, *87*, 774–781. [CrossRef]
4. Castro-Rodriguez, J.A.; Daszenies, C.; Garcia, M.; Meyer, R.; Gonzales, R. Adenovirus pneumonia in infants and factors for developing bronchiolitis obliterans: A 5-year follow-up. *Pediatr. Pulmonol.* **2014**, *41*, 947–953. [CrossRef]
5. Murdoch, D.R.; Morpeth, S.C.; Hammitt, L.L.; Driscoll, A.J.; Watson, N.L.; Baggett, H.C.; Brooks, W.A.; Knoll, M.D.; Feikin, D.R.; Kotloff, K.L.; et al. Microscopic analysis and quality assessment of induced sputum from children with Pneumonia in the PERCH study. *Clin. Infect. Dis.* **2017**, *64*, S271–S279. [CrossRef]
6. Quang, K.T.; Do, H.T.; Hung, V.P.; Vu, T.N.; Xuan, B.T.; Larsson, M.; Duong-Quy, S.; Nguyen-Thi-Dieu, T. Study on the co-infection of children with severe community-acquired pneumonia. *Pediatr. Int.* **2021**, *64*, e14853. [CrossRef]
7. KTran, Q.; Nguyen, P.M.; Tran, H.D.; Lu, H.Q.; La, V.G.T.; Nguyen, T. Lobar pneumonia and bacterial pathogens in Vietnamese children. *Curr. Pediatr. Res.* **2020**, *24*, 247–253.
8. Restrepo-Gualteros, S.M.; Jaramillo-Barberi, L.E.; Gonzalez-Santos, M.; Rodriguez-Martinez, C.E.; Perez, G.F.; Gutierrez, M.J. Characterization of cytomegalovirus lung infection in non-HIV infected children. *Viruses* **2014**, *6*, 2038–2051. [CrossRef]
9. Do, L.A.; Bryant, J.E.; Tran, A.T.; Nguyen, B.H.; Tran, T.T.; Tran, Q.H.; Vo, Q.B.; Tran Dac, N.A.; Trinh, H.N.; Nguyen, T.T.; et al. Respiratory Syncytial Virus and Other Viral Infections among Children under Two Years Old in Southern Vietnam 2009-2010: Clinical Characteristics and Disease Severity. *PLoS ONE* **2016**, *11*, e0160606. [CrossRef]
10. Tran, D.N.; Trinh, Q.D.; Pham, N.T.K.; Vu, M.P.; Ha, M.T.; Nguyen, T.Q.N.; Okitsu, S.; Hayakawa, S.; Mizuguchi, M.; Ushijima, H. Clinical and Epidemiological Characteristics of Acute Respiratory Virus Infections in Vietnamese Children. *Epidemiol. Infect.* **2016**, *144*, 527–536. [CrossRef]
11. Althouse, B.M.; Flasche, S.; Minh, L.N.; Thiem, V.D.; Hashizume, M.; Ariyoshi, K.; Anh, D.D.; Rodgers, G.L.; Klugman, K.P.; Hu, H.; et al. Seasonality of respiratory viruses causing hospitalizations for acute respiratory infections in children in Nha Trang, Vietnam. *Int. J. Infect. Dis.* **2018**, *75*, 18–25. [CrossRef]
12. Pacheco, G.A.; Gálvez, N.M.S.; Soto, J.A.; Andrade, C.A.; Kalergis, A.M. Bacterial and Viral Coinfections with the Human Respiratory Syncytial Virus. *Microorganisms* **2021**, *9*, 1293. [CrossRef]
13. William, J.; Barson, M. Community-Acquired Pneumonia in Children: Clinical Features and Diagnosis. 2022. Available online: www.uptodate.com (accessed on 14 November 2023).
14. Lim, L.M.; Woo, Y.Y.; De Bruyne, J.A.; Nathan, A.M.; Kee, S.Y.; Chan, Y.F.; Chiam, C.W.; Eg, K.P.; Thavagnanam, S.; Sam, I.C. Epidemiology, clinical presentation and respiratory sequelae of adenovirus pneumonia in children in Kuala Lumpur, Malaysia. *PLoS ONE* **2018**, *13*, e0205795. [CrossRef]
15. Zampoli, M.; Mukuddem-Sablay, Z. Adenovirus-associated pneumonia in South African children: Presentation, clinical course and outcome. *S. Afr. Med. J.* **2017**, *107*, 123–126. [CrossRef]
16. Hasegawa, J.; Mori, M.; Ohnishi, H.; Tsugawa, T.; Hori, T.; Yoto, Y.; Tsutsumi, H. Pneumococcal vaccination reduces the risk of community-acquired pneumonia in children. *Pediatr. Int.* **2017**, *59*, 316–320. [CrossRef]
17. Korppi, M. Antibiotic therapy in children with community-acquired pneumonia. *Acta Paediatr.* **2021**, *110*, 3246–3250. [CrossRef]
18. Le Roux, D.M.; Nicol, M.P.; Vanker, A.; Nduru, P.M.; Zar, H.J. Factors associated with serious outcomes of pneumonia among children in a birth cohort in South Africa. *PLoS ONE* **2021**, *16*, e0255790. [CrossRef]
19. Zhang, L.; Silva, F. Bronchiolitis obliterans in children. *J. Pediatr.* **2000**, *76*, 185–192. [CrossRef]
20. Zhong, L.; Lin, J.; Dai, J. Risk factors for the development of bronchiolitis obliterans in children with severe adenovirus pneumonia: A retrospective study with dose-response analysis. *J. Med. Virol.* **2020**, *92*, 3093–3099. [CrossRef]
21. British Thoracic Society Standards of Care Committee. British Thoracic Society Guidelines for the Management of Community Acquired Pneumonia in Childhood. *Thorax* **2002**, *57* (Suppl. 1), i1–i24. [CrossRef]
22. Virkki, R.; Juven, T.; Rikalainen, H.; Svedström, E.; Mertsola, J.; Ruuskanen, O. Differentiation of bacterial and viral pneumonia in children. *Thorax* **2002**, *57*, 438–441. [CrossRef]
23. UNICEF. Save the Children, and Every Breath Counts. Every Child's Right to Survive: A 2020 Agenda to End Pneumonia Deaths. 2020. Available online: https://www.unicef.org/reports/every-childs-right-survive-pneumonia (accessed on 14 November 2023).

viruses

Systematic Review

Impact of Nonpharmaceutical Interventions during the COVID-19 Pandemic on the Prevalence of Respiratory Syncytial Virus in Hospitalized Children with Lower Respiratory Tract Infections: A Systematic Review and Meta-Analysis

José J. Leija-Martínez [1], Luis A. Esparza-Miranda [1,2], Gerardo Rivera-Alfaro [1] and Daniel E. Noyola [1,2,*]

1 Research Center in Health Sciences and Biomedicine (CICSaB), Universidad Autónoma de San Luis Potosí, San Luis Potosí 78210, Mexico; jesus.leija@uaslp.mx (J.J.L.-M.); a250764@alumnos.uaslp.mx (L.A.E.-M.); a238871@alumnos.uaslp.mx (G.R.-A.)

2 Microbiology Department, Facultad de Medicina, Universidad Autónoma de San Luis Potosí, San Luis Potosí 78210, Mexico

* Correspondence: dnoyola@uaslp.mx

Abstract: During the COVID-19 pandemic, nonpharmaceutical interventions (NPIs) were implemented in order to control the transmission of SARS-CoV-2, potentially affecting the prevalence of respiratory syncytial virus (RSV). This review evaluated the impact of NPIs on RSV-related hospitalizations in children during the lockdown (2020–2021) compared to the pre-pandemic (2015–2020) and post-lockdown (2021–2022) periods. In this systematic review and meta-analysis, we searched through PubMed, Scopus, and Web of Science for studies published in English between 1 January 2015 and 31 December 2022. Additionally, we conducted hand searches of other records published between 1 January 2023 and 22 January 2024. Our target population was hospitalized children aged 0–18 years with RSV-related lower respiratory tract infections confirmed through immunofluorescence, antigen testing, or molecular assays. We focused on peer-reviewed observational studies, analyzing the primary outcome of pooled RSV prevalence. A generalized linear mixed model with a random-effects model was utilized to pool each RSV prevalence. Heterogeneity was assessed using Cochran's Q and I^2 statistics, while publication bias was evaluated through funnel plots and Egger's tests. We identified and analyzed 5815 publications and included 112 studies with 308,985 participants. Notably, RSV prevalence was significantly lower during the lockdown period (5.03% [95% CI: 2.67; 9.28]) than during the pre-pandemic period (25.60% [95% CI: 22.57; 28.88], $p < 0.0001$). However, RSV prevalence increased notably in the post-lockdown period after the relaxation of COVID-19 prevention measures (42.02% [95% CI: 31.49; 53.33] vs. 5.03% [95% CI: 2.67; 9.28], $p < 0.0001$). Most pooled effect estimates exhibited significant heterogeneity (I^2: 91.2% to 99.3%). Our findings emphasize the effectiveness of NPIs in reducing RSV transmission. NPIs should be considered significant public health measures to address RSV outbreaks.

Keywords: respiratory syncytial virus; nonpharmaceutical interventions; lockdown; COVID-19; SARS-CoV-2; respiratory tract infection

Citation: Leija-Martínez, J.J.; Esparza-Miranda, L.A.; Rivera-Alfaro, G.; Noyola, D.E. Impact of Nonpharmaceutical Interventions during the COVID-19 Pandemic on the Prevalence of Respiratory Syncytial Virus in Hospitalized Children with Lower Respiratory Tract Infections: A Systematic Review and Meta-Analysis. *Viruses* **2024**, *16*, 429. https://doi.org/10.3390/v16030429

Academic Editors: Emanuele Castagno and Irene Raffaldi

Received: 30 December 2023
Revised: 15 February 2024
Accepted: 19 February 2024
Published: 11 March 2024

1. Introduction

Respiratory syncytial virus (RSV) is the primary cause of early childhood lower respiratory tract infections (LRTIs), leading to severe illness and high mortality rates [1]. In 2019, RSV resulted in 33 million LRTI episodes, 3.6 million hospitalizations, and 101,400 deaths among children aged 0–60 months [1]. This virus exhibits seasonal transmission patterns, with epidemic peaks occurring in autumn and winter in temperate climates [2,3].

The declaration of coronavirus disease 2019 (COVID-19) as a global public health emergency by the World Health Organization (WHO) prompted the adoption of nonpharmaceutical interventions (NPIs) aimed at mitigating the transmission of severe acute respiratory syndrome coronavirus 2 (SARS-CoV-2) [4]. These interventions likely altered the seasonality patterns of RSV and profoundly impacted its prevalence among hospitalized children with LRTIs, particularly during the 2020–2021 season [5–7].

Ravkin et al. [7] utilized Google Trends search volumes as an indicator of viral circulation and observed a disruption between the peak latency and magnitude of RSV during the pandemic. This observed phenomenon could be attributed to NPIs, emphasizing the significant impact of population mobility on RSV incidence. Multiple countries have reported a substantial decrease in RSV-related LRTI hospitalizations, with some indicating a 90–99% reduction [5,6]. For instance, in England, Bardsley et al. [5] reported a dramatic fall in RSV test positivity through PCR of 99.6% among children under five years old, as documented by the Respiratory DataMart System (RDS). Remarkably, in Italy, Pruccoli et al. [6] reported only three cases of RSV-related hospital admissions among children under three years old across fifteen pediatric hospitals. As a result, the 2020–2021 RSV season presented a real-world opportunity to assess the effectiveness of NPIs in reducing RSV transmission.

Due to the similarities in transmission mechanisms between RSV and SARS-CoV-2, the global implementation of NPIs resulted in a decline in RSV infections [8]. However, following the relaxation of COVID-19 restrictions, RSV seasonality patterns and outbreaks have returned to pre-pandemic levels or even increased [5,9–12].

Presumably, NPIs may have influenced RSV-related hospitalizations, underscoring the importance of comprehensively examining their effects. To address this concern, this systematic review aimed to assess the impact of NPIs on the prevalence of RSV among hospitalized children with LRTIs during the early pandemic (lockdown) period (2020–2021) in comparison to the pre-pandemic (2015–2020) and post-lockdown periods (2021–2022). Of note, because winter seasons start in the second half of a given year and end during the first half of a given year, the periods appear to include a "repeated" year, but cases were assigned only to one period, based on the months of the year.

2. Materials and Methods

2.1. Search Strategy and Selection Criteria

The systematic review protocol was registered on the International Prospective Register of Systematic Reviews (PROSPERO) with the registration number CRD42022376951 and adhered to the Preferred Reporting Items for Systematic Reviews and Meta-Analyses (PRISMA) statement, published in 2020 [13,14]. For detailed information, please refer to Table S1 in the Supplementary Material SA for the PRISMA 2020 Checklist. The PICOS strategy was used to establish the eligibility criteria (population, intervention, comparison, outcome, and study design) [15]. Our research question was as follows: "What is the impact of NPIs during the COVID-19 pandemic on the prevalence of RSV-related LRTI hospitalizations in children?"

This study examined children aged 0–18 years hospitalized with RSV-related LRTIs, excluding nosocomial infections. RSV diagnosis was confirmed using immunofluorescence, antigen testing, or molecular assays. The prevalence of RSV hospitalizations was assessed for each winter season; studies in which pooled data from several seasons were reported were excluded. Peer-reviewed observational studies (cohort, case–control, and cross-sectional) were considered, with an RSV season defined as a six-month local epidemic. Case reports, reviews, editorials, and duplicates were excluded.

A comprehensive search through the PubMed, Scopus, and Web of Science databases was conducted to identify relevant articles published between 1 January 2015 and 31 December 2022. Additionally, other methods involving hand searching were carried out for articles published from 1 January 2023 to 22 January 2024. We exclusively incorporated studies conducted in the English language. Table S2 in the Supplementary Material SA includes the search terms employed for each database. Additionally, we thoroughly examined the bibliographies of pertinent research articles. The study selection process involved LAEM and JJLM independently screening records, reviewing full reports, and discussing articles. A third reviewer (DEN) made the final decision if disagreements arose. Pilot screening refined the process before the actual screening.

Two reviewers (LAEM and GRA) extracted data using Microsoft® Excel 365 spreadsheets, conducting pilot extraction to ensure consistency. Cross-checking was conducted to ensure accurate extraction, and a final spreadsheet was obtained. A third reviewer (DEN) resolved any disagreements. Researchers emailed the corresponding author if additional data were needed for inclusion or clarification. The following items were collected: RSV prevalence, period, seasons, WHO region, age, sex, study design, timing of data collection, diagnostic technique, and sample type.

The primary outcome was the pooled RSV prevalence in hospitalized children with LRTIs across the pre-pandemic, lockdown, and post-lockdown periods. As a secondary outcome, we examined the prevalence of intensive care unit (ICU) admissions and deaths among children with LRTIs related to RSV during these three periods. The tool developed by Hoy et al. [16] for prevalence studies was employed to assess the risk of bias, categorizing it as low (8–10), moderate (5–7), or high (0–4) (Table S3 in Supplementary Material SA). The reviewers LAEM and JJLM independently analyzed and cross-checked bias assessments, resolving any disagreements through discussion with a third reviewer (DEN). The quality of evidence was assessed using the AMSTAR 2 instrument (A MeaSurement Tool to Assess systematic Reviews 2) [17].

2.2. Data Analysis

We summarized the study characteristics in tables and narrative synthesis. We determined the proportion of RSV-related LRTI cases by dividing the number of RSV-positive samples by the total number of samples tested. Then, we transformed the proportion into a prevalence by multiplying it by 100.

Our meta-analysis aimed to estimate and report the pooled prevalence with a 95% confidence interval (CI). We observed a wide prevalence range, varying from nearly 0% to 100%. This wide range could lead to overestimating precision due to significant variance. To address this issue, we performed a generalized linear mixed model using a maximum-likelihood estimator for τ^2 and a random-effects model to pool each RSV prevalence [18]. Finally, the results were presented through forest plots, tables, and narratives to display the pooled effect estimates.

Heterogeneity was assessed visually through forest plots and using Cochran's Q and I^2 statistics. Cochran's Q with $p < 0.1$ indicated significant heterogeneity among studies [15,19]. The I^2 statistic categorized heterogeneity as follows: (a) <25%: low heterogeneity; (b) 25–50%: moderate heterogeneity; (c) ≥50%: high heterogeneity [15,19]. A random effects model was used due to the observed high heterogeneity. Funnel plots and Egger's tests assessed potential publication bias, with a p-value < 0.1 being considered statistically significant. Analyses were performed using the RStudio (R version 4.2.3) "meta" package (version 6.2-1), with a p-value < 0.05 indicating statistical significance.

Subgroup, sensitivity, and meta-regression analyses explored sources of heterogeneity based on study and participant characteristics: period, seasons, WHO region, age, study design, timing of data collection, risk of bias, diagnostic technique, and sample type.

Because of the great variety of diagnostic techniques used, we grouped antigen tests and direct immunofluorescence into a single category called immune assays. Additionally, all types of PCR, including multiplex PCR, multiplex RT–PCR, PCR, qPCR, RT–PCR, or RT–qPCR, were grouped into a single category known as molecular assays.

Sensitivity analyses included only studies using molecular assays for RSV diagnosis. Furthermore, we conducted a stratified analysis based on age strata (<2, <5, <10, <15, and <18 years), WHO regions, and countries to compare the pandemic (lockdown, 2020–2021), pre-pandemic (2015–2020), and post-lockdown (2021–2022) periods. A bubble plot illustrates the relationship between RSV prevalence and the study participants' average age (in months). In the meta-regression model, the presence of residual heterogeneity in RSV prevalence was suggested by the explained heterogeneity (R^2).

3. Results

3.1. Search Results

A total of 8091 records were identified from the PubMed (n = 4367), Scopus (n = 1191), and Web of Science (n = 2410) databases, along with hand searching (n = 123). After removing duplicate records (n = 2276), we screened 5815 unique records (Figure 1). Among these, we assessed 578 full reports for eligibility, excluding 466 reports due to them not meeting the inclusion criteria. Ultimately, we included 112 studies in this systematic review and meta-analysis. For the list of the 466 full reports that were excluded and the reasons for their exclusion, please see Supplementary Material SB.

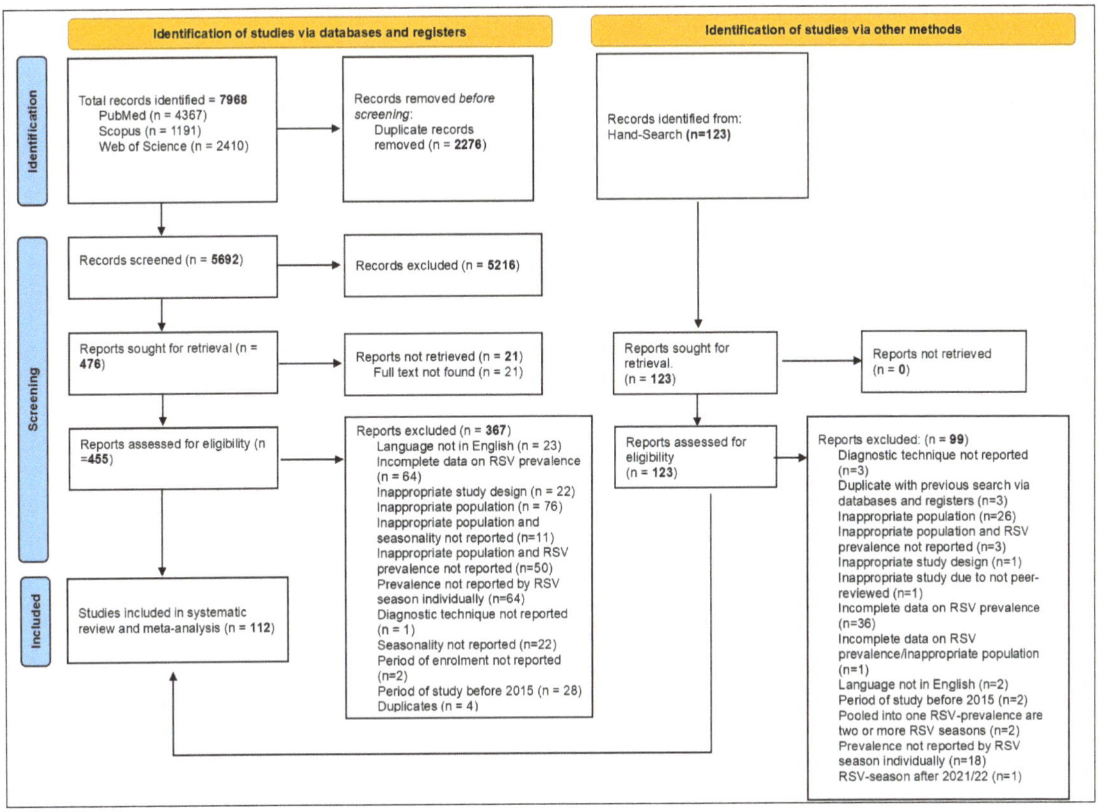

Figure 1. Study selection.

3.2. Characteristics of the Included Articles

As illustrated in Table 1, most studies in this review focused on children under two years of age (35/112; 31.3%). All participants were recruited between January 2015 and December 2022. Cross-sectional studies were more common (75/112; 67.0%), and most employed a prospective recruitment strategy (58/112; 51.8%). Notably, the most commonly used sampling method was consecutive sampling (110/112; 98.2%), and most studies exhibited a low risk of bias (71/112; 63.4%). For a comprehensive analysis of the risk of bias in each study, please refer to Tables S3 and S4 in the Supplementary Material SA.

Table 1. Sociodemographic and clinical characteristics of the included studies.

Characteristics	N = 112	%
Age range		
• <2 years	35	31.3
• <5 years	22	19.6
• <10 years	4	3.6
• <15 years	20	17.9
• <18 years	27	24.1
• Not reported	4	3.6
Male range (%)	39.1–73.8%	
Period of inclusion of participants; range	January 2015–December 2022	
Year of publication; range	2017–2023	
Study design		
• Cross-sectional	75	67
• Longitudinal	37	33
Sampling method		
• Consecutive	110	98.2
• Random	2	1.8
Timing of data collection		
• Ambispective	4	3.6
• Prospective	58	51.8
• Retrospective	50	44.6
Study bias		
• Low risk	71	63.4
• Moderate risk	41	36.6
WHO region		
• African	7	6.3
• Americas	13	11.6
• Eastern Mediterranean	3	2.7
• European	47	42
• South-East Asia	8	7.1
• Western Pacific	34	30.4
Type of assay		
• Immune assays ¥	22	19.6
• Molecular assays ¥¥	85	75.9
• Mixed assays ¥¥¥	5	4.5
Diagnostic technique $		
• Antigen	6	5.4
• Direct immunofluorescence	16	14.3
• Mixed ¥¥¥	5	4.5
• Multiplex PCR	14	12.5
• Multiplex RT–PCR	18	16.1
• PCR	16	14.3
• qPCR	2	1.8
• RT–PCR	31	27.7
• RT–qPCR	4	3.6
Sample type		
• Mixed *	7	6.3
• Nasal secretions	9	8
• Nasopharyngeal secretions	62	55.4
• Nasopharyngeal secretions or BLF	11	9.8
• Not reported	8	7.1
• Oropharyngeal swab	7	6.3
• Sputum	3	2.7
• Throat swab	5	4.5

¥ Immune assays: antigen testing or direct immunofluorescence. ¥¥ Molecular assays: multiplex PCR, multiplex RT–PCR, PCR, qPCR, RT–PCR, or RT–qPCR. ¥¥¥ Mixed assays: antigen testing/RT–PCR, direct immunofluorescence/RT–qPCR/antigen testing, direct immunofluorescence/RT–PCR, or indirect immunofluorescence/PCR. $ The indicated diagnostic technique is according to the authors' description. * Mixed specimens: nasopharyngeal secretions and nasal secretions or nasal and throat secretions. Abbreviations: WHO, World Health Organization; BLF, bronchoalveolar lavage fluid.

Across all six WHO regions, our review encompassed studies primarily reported from the European region (47/112; 42.0%), followed by the Western Pacific region (34/112; 30.4%). It is worth noting that these two regions accounted for 72.3% (81/112) of all included studies. Furthermore, among the 37 countries that provided the 112 studies in this review, a large proportion of studies were conducted in China (27/112; 24.1%) and Italy (20/112; 17.9%) (Table S5 in the Supplementary Material SA). Molecular assays, particularly RT–PCR and nasopharyngeal secretions, were the most frequently utilized diagnostic technique and type of sample, respectively (31/112; 27.7% and 62/112; 55.4%, respectively). For a comprehensive analysis of each study's characteristics with detailed references, please refer to Tables S6 and S7 in the Supplementary Material SA.

The 112 included studies provide 221 RSV prevalences, which varied significantly across all studies, ranging from 0% to 82.24%. During the lockdown period, the proportions ranged from 0% to 77.78%. Notably, ten studies reported proportions of 0%, while two studies observed no occurrences of RSV-positive tests or hospitalizations for LRTIs during this lockdown period [6,20–25]. Of note, in some studies, such as the report by Stera et al. [25] and Pruccoli et al. [6], the numerator and denominator were zero during the 2020/21 season, making it mathematically implausible to calculate a proportion; therefore, they were excluded from the meta-analysis. The proportion ranges in the pre-pandemic and post-lockdown periods were 4.65% to 79.31% and 4.76% to 82.24%, respectively. Please refer to Figure S1 in the Supplementary Material SA for a comprehensive breakdown of the 221 RSV prevalences based on each study and RSV season.

This review incorporated 112 studies encompassing 308,985 participants from 37 countries, yielding a total of 221 prevalences. The overall pooled prevalence of RSV was 21.51% [95% CI: 18.42; 24.96]. Notably, there was significant heterogeneity in the effect size (Q-value = 22,893.65, $I^2 = 99.0\%$, $p < 0.0001$) (Figure 2 and Figure S1 in the Supplementary Material SA).

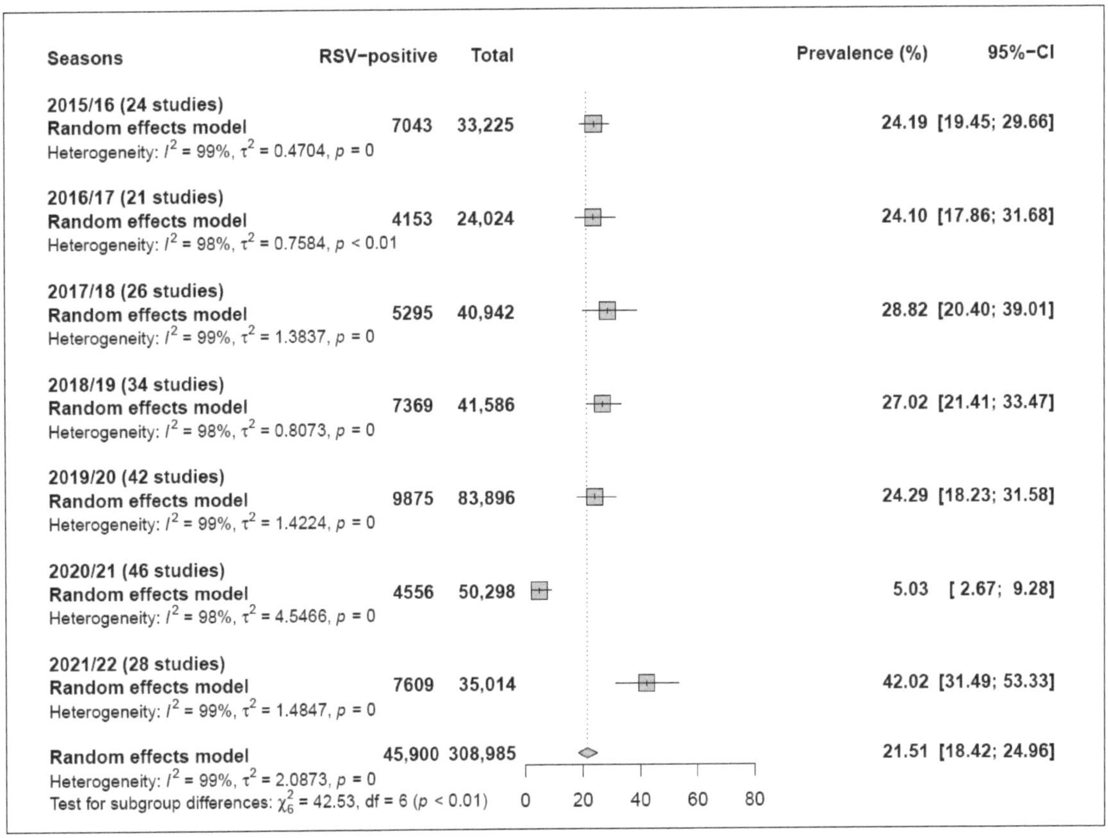

Figure 2. Respiratory syncytial virus (RSV) prevalences according to RSV season. For all 221 prevalences and their respective references, please refer to Figure S1 and the reference section at the end of the Supplementary Material SA (pages 39–59).

3.3. Subgroup Analysis

Regarding the subgroup analysis, the prevalence of RSV was significantly lower during the lockdown period than during the pre-pandemic period (5.03% [95% CI: 2.67; 9.28] vs. 25.6% [95% CI: 22.57; 28.88], $p < 0.0001$). Interestingly, the prevalence of RSV increased in the post-lockdown period after relaxing COVID-19 mitigation measures compared to the lockdown period (42.02% [95% CI: 31.49; 53.33] vs. 5.03% [95% CI: 2.67; 9.28], $p < 0.0001$). A detailed subgroup analysis is provided in Table 2.

Comparisons of RSV prevalence among pre-pandemic seasons (2015–2016 through 2019–2020) showed no difference ($p = 0.8651$). However, when each pre-pandemic season was compared individually with the lockdown period (2020–2021) and the post-lockdown season (2021–2022), similar results were observed as when all pre-pandemic seasons were compared as a single group: a lower prevalence during the lockdown period and a higher prevalence during the post-lockdown period ($p < 0.0001$) (Table 2 and Figure 2).

Overall, the European region exhibited a higher RSV prevalence than the other WHO regions ($p = 0.0004$), and children under two years of age had a higher prevalence than the other age groups ($p < 0.0001$). On the other hand, there was no significant difference in RSV prevalence between studies with low and moderate risk of bias ($p = 0.1453$) (Table 2). Noticeably, studies using molecular assays demonstrated a significantly higher prevalence than those based on immune assays (24.96% [95% CI: 20.77; 29.69] vs. 13.94% [95% CI: 10.65; 18.04], $p = 0.0007$). Specifically, the qPCR test showed the highest RSV prevalence among all diagnostic techniques ($p = 0.0016$). Moreover, nasopharyngeal secretions and sputum specimens were associated with a higher RSV prevalence than other specimen types ($p = 0.0023$) (Table 2).

Most pooled effect estimates exhibit substantial heterogeneity, with I^2 statistics ranging from 91.2% to 99.3%. Publication bias was detected for the overall pooled prevalence of RSV, as indicated by the funnel plot and corroborated through Egger's test ($p < 0.0001$) (Table 2 and Figure 3).

Nevertheless, it is worth highlighting that there was no publication bias during the lockdown period, as indicated by Egger's test ($p = 0.5569$) (Table 2). It is also important to note that most subgroup analyses found publication bias when analyzed using Egger's test, except for the type of assay and diagnostic technique.

3.4. Sensitivity Analysis

We conducted a sensitivity analysis explicitly focusing on the impact of molecular assay diagnostic techniques on RSV proportions. The overall pooled prevalence using molecular assays was 24.96% [95% CI: 20.77; 29.69], which showed a 3.46% increase compared to the meta-analysis that included all diagnostic techniques. Similar to the initial meta-analysis, evidence of publication bias was found, demonstrated by the asymmetry in the funnel plot and corroborated by Egger's test ($p = 0.0350$) (Table S8 and Figure S2 in the Supplementary Material SA). Nonetheless, this sensitivity analysis reported no publication bias during the lockdown (2020/21) period based on Egger's test ($p = 0.2778$). The pooled RSV prevalence during the lockdown period was significantly lower than that in the pre-pandemic period (3.82% [95% CI: 1.53; 9.22] vs. 30.45% [95% CI: 26.68; 34.49], $p < 0.0001$). Interestingly, there was a significant increase in the pooled RSV prevalence during the post-lockdown period compared to the lockdown period (44.29% [95% CI: 32.61; 56.63] vs. 3.82% [95% CI: 1.53; 9.22], $p < 0.0001$) (Table S8 in Supplementary Material SA). Remarkably, the findings from the sensitivity analysis regarding RSV prevalence across different periods and seasons align with those of the initial meta-analysis.

Table 2. Global and subgroup pooled prevalence of respiratory syncytial virus in children with lower respiratory tract infections.

Groups	Studies (n)	RSV-Positive (n)	Total (n)	Pooled Prevalence (95% CI)	Q-Value	I^2 (%)	*p* Value Heterogeneity	*p* Value Egger Test	*p* Value Subgroup Difference
Overall	112	45,900	308,985	21.51 [18.42; 24.96]	22,893.65	99.0	<0.0001	<0.0001	NA
Subgroup analyses									
Period									<0.0001
Pre-pandemic (2015/20)	83	33,735	223,673	25.60 [22.57; 28.88]	15,032.9	99.0	<0.0001	<0.0001	
Lockdown (2020/21)	46	4556	50,298	5.03 [2.67; 9.28]	2581.79	98.3	<0.0001	0.5569	
Post-lockdown (2021/22)	28	7609	35,014	42.02 [31.49; 53.33]	3765.28	99.3	<0.0001	<0.0001	
Seasons									<0.0001
2015/16	24	7043	33,225	24.19 [19.45; 29.66]	1620.84	98.60	<0.0001	0.2232	
2016/17	21	4153	24,024	24.10 [17.86; 31.68]	1298.39	98.50	<0.0001	0.035	
2017/18	26	5295	40,942	28.82 [20.40; 39.01]	2855.18	99.10	<0.0001	0.0003	
2018/19	34	7369	41,586	27.02 [21.41; 33.47]	2139.94	98.50	<0.0001	0.0011	
2019/20	42	9875	83,896	24.29 [18.23; 31.58]	5676.49	99.30	<0.0001	0.0003	
2020/21	46	4556	50,298	5.03 [2.67; 9.28]	2581.79	98.30	<0.0001	0.5569	
2021/22	28	7609	35,014	42.02 [31.49; 53.33]	3765.28	99.30	<0.0001	<0.0001	
WHO Region									0.0004
African	7	837	3602	19.42 [14.19; 25.98]	224.25	96.0	<0.0001	0.0276	
Americas	13	2029	6370	11.17 [2.19; 41.42]	194.76	93.3	<0.0001	0.7522	
Eastern Mediterranean	3	151	568	26.53 [15.66; 41.25]	33.53	94.0	<0.0001	NA	
European	47	14,877	70,357	28.90 [22.73; 35.98]	7570.45	98.7	<0.0001	0.0004	
South–East Asia	8	1283	5987	24.85 [14.25; 39.67]	758.49	98.7	<0.0001	0.6948	
Western Pacific	34	26,723	222,101	14.71 [12.28; 17.52]	8245.53	99.0	<0.0001	0.1071	
Age									<0.0001
<2 years	35	6815	17,696	43.54 [35.56; 51.87]	1972.43	96.7	<0.0001	0.0017	
<5 years	22	7223	36,599	25.59 [21.03; 30.76]	2760.33	98.8	<0.0001	0.0781	
<10 years	4	7706	65,262	17.34 [11.89; 24.58]	1267.48	99.0	<0.0001	0.1347	
<15 years	20	5745	40,646	17.58 [12.82; 23.64]	4016.97	99.1	<0.0001	0.7428	
<18 years	27	17,408	139,980	10.17 [7.29; 14.02]	5,380.77	99.0	<0.0001	0.9850	
Not reported	4	1003	8802	9.94 [5.19; 18.18]	276.85	95.7	<0.0001	0.3230	
Design									0.0840
Cross-sectional	75	35,091	263,251	19.64 [16.55; 23.15]	15,271.27	98.9	<0.0001	<0.0001	
Longitudinal	37	10,809	45,734	27.34 [19.47; 36.94]	4589.11	98.8	<0.0001	0.0379	

Table 2. *Cont.*

Groups	Studies (n)	RSV-Positive (n)	Total (n)	Pooled Prevalence (95% CI)	Q-Value	I^2 (%)	*p* Value Heterogeneity	*p* Value Egger Test	*p* Value Subgroup Difference
Timing of data collection									0.5983
Ambispective	4	1449	11,714	15.30 [4.70; 39.82]	657.29	98.8	<0.0001	NA	
Prospective	58	12,676	66,479	23.44 [18.63; 29.05]	7496.05	98.9	<0.0001	0.1370	
Retrospective	50	31,775	230,792	20.82 [16.90; 25.37]	12,692.16	99.0	<0.0001	0.0007	
Risk of bias									0.1453
Low risk	71	31,179	222,366	23.29 [19.18; 27.98]	17,585.49	99.2	<0.0001	<0.0001	
Moderate risk	41	14,721	86,619	18.49 [14.32; 23.54]	5058.52	98.5	<0.0001	0.0752	
Type of assay									0.0007
Immune assays ¥	22	19,476	180,647	13.94 [10.65; 18.04]	4163.17	98.9	<0.0001	0.5255	
Molecular assays ¥¥	85	22,523	103,157	24.96 [20.77; 29.69]	10,441.27	98.5	<0.0001	0.0350	
Mixed assays ¥¥¥	5	3901	25,181	14.74 [8.27; 24.91]	872.62	96.3	<0.0001	0.3155	
Diagnostic technique $									0.0016
Antigen testing	6	8018	75,112	17.78 [10.71; 28.05]	1187.05	98.9	<0.0001	0.4242	
Direct immunofluorescence	16	11,458	105,535	12.46 [9.14; 16.77]	2892.56	99.0	<0.0001	0.8831	
Mixed assays ¥¥¥	5	3901	25,181	14.74 [8.27; 24.91]	1187.05	98.9	<0.0001	0.3155	
Multiplex PCR	14	2083	10,831	27.32 [14.57; 45.30]	2965.49	99.0	<0.0001	0.0439	
Multiplex RT–PCR	18	2215	7385	30.89 [22.03; 41.42]	872.62	98.3	<0.0001	0.3490	
PCR	16	7247	32,319	24.91 [16.89; 35.12]	1756.55	98.7	<0.0001	0.8591	
qPCR	2	52	140	38.93 [21.17; 60.20]	828.23	97.0	0.0008	NA	
RT–PCR	31	9193	46,668	22.15 [16.57; 28.95]	3648.00	99.1	<0.0001	0.0936	
RT–qPCR	4	1733	5814	17.82 [3.59; 55.79]	11.32	91.2	<0.0001	NA	
Sample type									0.0023
Mixed specimens *	7	3091	14,765	18.35 [8.59; 34.98]	1408.90	99.0	<0.0001	0.7840	
Nasal secretions	9	1209	3493	21.39 [6.94; 49.83]	263.44	95.8	<0.0001	0.5521	
Nasopharyngeal secretions	62	21,542	114,495	26.25 [21.48; 31.66]	9150.13	98.7	<0.0001	<0.0001	
Nasopharyngeal secretions or BLF	11	4374	52,107	12.12 [8.34; 17.28]	1848.89	98.6	<0.0001	0.0770	
Not reported	8	10,981	85,775	21.26 [10.99; 37.13]	2581.81	99.3	<0.0001	0.0036	
Oropharyngeal secretions	7	2663	20,375	15.48 [12.23; 19.39]	778.85	97.7	<0.0001	0.2721	
Sputum	3	549	2063	25.23 [18.38; 33.58]	44.03	95.5	<0.0001	NA	
Throat secretions	5	1491	15,912	16.79 [7.88; 32.26]	1097.90	99.3	<0.0001	NA	

¥ Immune assays: Antigen testing; or Direct immunofluorescence. ¥¥ Molecular assays: Multiplex PCR, Multiplex RT–PCR, PCR, qPCR, RT–PCR, or RT–qPCR. ¥¥¥ Mixed assays: Antigen testing/RT–PCR; direct immunofluorescence/RT–qPCR/antigen testing; direct immunofluorescence/RT–PCR; or indirect immunofluorescence/PCR. $ The indicated diagnostic technique is according to what is referred to by the authors. * Mixed specimens: nasopharyngeal secretions and nasal secretions or nasal and throat secretions. Abbreviations: WHO, World Health Organization; BLF, bronchoalveolar lavage fluid; NA, not applicable.

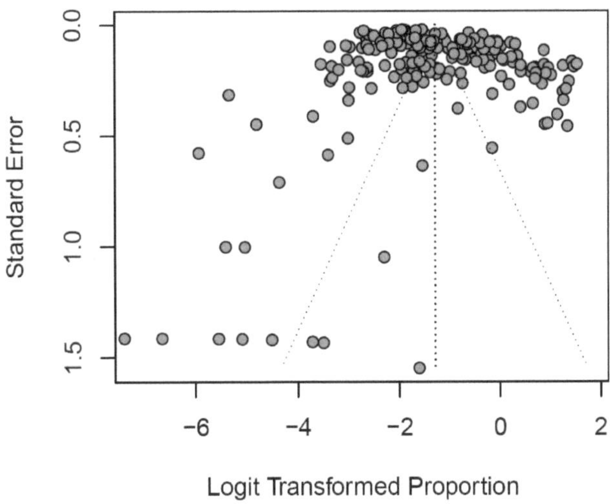

Figure 3. Funnel plot of overall pooled prevalence of respiratory syncytial virus in lower respiratory tract infections.

3.5. Stratified Analysis by World Health Organizations Regions, Countries, and Age Stratum

As shown in Table 1, the vast majority of WHO regions and countries in this review were from the Western Pacific and Europe, with China and Italy representing their respective regions. Additionally, a high prevalence of RSV was observed in Europe, as reported in subgroup and meta-regression model analyses (Table 2 and Table S9 in the Supplementary Material SA).

In Europe, the impact of NPIs was evident; during the lockdown period, the pooled prevalence of RSV was significantly lower than during the pre-pandemic period (4.89% [95% CI: 1.68; 13.39] vs. 34.81% [95% CI: 29.47; 40.56], $p < 0.0001$). In contrast, there was a notable upward trend in RSV prevalence in the post-lockdown period after relaxing COVID-19 mitigation measures compared to the lockdown period (55.18% [95% CI: 42.96; 66.80] vs. 4.89% [95% CI: 1.68; 13.39], $p < 0.0001$) (Table 3).

Similar to the European region, Italy exhibited a low RSV prevalence during the lockdown period compared to the pre-pandemic period (5.93% [95% CI: 0.89; 30.68] vs. 51.87% [95% CI: 43.11; 60.51], $p = 0.0024$). After relaxing COVID-19 measures, Italy experienced a significant upward trend in RSV prevalence (63.59% [95% CI: 53.25; 72.81] vs. 5.93% [95% CI: 0.89; 30.68], $p = 0.0024$) (Table 3). On the other hand, though less pronounced, the Western Pacific region and China demonstrated similar trends in RSV prevalences during the three periods, showing a decline in the lockdown period and an increase in the post-lockdown period ($p = 0.0002$) (Table 3). It is worth underscoring that we were unable to perform these meta-analyses for the remaining WHO regions and countries due to the lack of enough studies for all three periods.

Moreover, as evident from subgroup and meta-regression model analyses, age plays a crucial role in RSV prevalence, especially among infants and toddlers under two years of age (Table 2 and Table S9 in the Supplementary Material SA). The effect of NPIs was more significant in the stratum under two years old. RSV prevalence during the lockdown period was lower than in the pre-pandemic period (6.46% [95% CI: 1.19; 28.29] vs. 47.82% [95% CI: 42.06; 53.65], $p < 0.0001$) (Table 4). Meanwhile, a major increase in RSV prevalence was observed in the post-lockdown period (67.61% [95% CI: 57.01; 76.67] vs. 6.46% [95% CI: 1.19; 28.29], $p = 0.0024$). The rest of the age strata showed similar trends in RSV prevalence during the pre-pandemic, lockdown, and post-lockdown periods as observed in the under two years old stratum; however, this was less marked (Table 4).

Table 3. Analysis by World Health Organization regions and countries.

Groups	Studies (n)	RSV-Positive (n)	Total (n)	Pooled Prevalence (95% CI)	Q-Value	I^2 (%)	*p* Value Heterogeneity	*p* Value Egger Test	*p* Value Subgroup Difference
WHO Region ##									
European	47	14,877	70,357	28.90 [22.73; 35.98]	7570.45	98.7	<0.0001	0.0004	
Period									<0.0001
Pre-pandemic (2015/20)	28	10,348	48,140	34.81 [29.47; 40.56]	3905.86	98.50	<0.0001	0.0001	
Lockdown (2020/21)	23	1028	12,898	4.89 [1.68; 13.39]	853.67	97.40	<0.0001	0.7463	
Post-lockdown (2021/22)	18	3501	9319	55.18 [42.96; 66.80]	1479.62	98.90	<0.0001	0.0093	
Western Pacific	34	26,723	222,101	14.71 [12.28; 17.52]	8245.53	99.0	<0.0001	0.1071	
Period									0.0024
Pre-pandemic (2015/20)	29	19,728	162,313	16.12 [13.28; 19.43]	6290.45	99.10	<0.0001	0.0424	
Lockdown (2020/21)	16	3215	34,966	8.03 [4.99; 12.68]	1240.05	98.80	<0.0001	0.5502	
Post-lockdown (2021/22)	8	3780	24,822	22.83 [15.42; 32.42]	356.45	98.00	<0.0001	NA	
Countries &&									
Italy	20	5437	14,928	47.14 [37.07; 57.45]	1877.72	97.8	<0.0001	0.0011	
Period									0.0024
Pre-pandemic (2015/20)	10	2955	9600	51.87 [43.11; 60.51]	815.15	97.40	<0.0001	<0.0001	
Lockdown (2020/21)	8	43	801	5.93 [0.89; 30.68]	110.71	93.70	<0.0001	NA	
Post-lockdown (2021/22)	13	2439	4527	63.59 [53.25; 72.81]	381.83	96.90	<0.0001	0.0136	
China	27	23,416	211,831	11.33 [9.61; 13.30]	4416.56	98.6	<0.0001	0.6889	
Period									0.0002
Pre-pandemic (2015/20)	23	16,898	153,351	12.04 [10.12; 14.26]	3350.33	98.70	<0.0001	0.79	
Lockdown (2020/21)	12	2868	34,012	6.64 [4.83; 9.07]	347.18	96.80	<0.0001	0.0314	
Post-lockdown (2021/22)	6	3650	24,468	19.89 [12.49; 30.15]	237.17	97.90	<0.0001	NA	

The World Health Organization (WHO) regions, including Africa, the Americas, the Eastern Mediterranean, and South East Asia, lack sufficient prevalence data for conducting this meta-analysis across periods. && Furthermore, the remaining countries also lack adequate prevalence data for conducting this meta-analysis across periods. Abbreviations: NA, not applicable.

Table 4. Analysis by age stratum.

Groups	Studies (n)	RSV-positive (n)	Total (n)	Pooled Prevalence (95%CI)	Q-Value	I^2 (%)	*p* Value Heterogeneity	*p* Value Egger Test	*p* Value Subgroup Difference
Age stratum									
Age < 2 years	35	6815	17,696	43.54 [35.56; 51.87]	1972.43	96.7	<0.0001	0.0017	NA
Period									<0.0001
Pre-pandemic (2015/20)	23	4844	12,939	47.82 [42.06; 53.65]	1019.71	96.10	<0.0001	<0.0001	
Lockdown (2020/21)	14	261	1359	6.46 [1.19; 28.29]	137.86	90.60	<0.0001	0.1192	
Post-lockdown (2021/22)	12	1710	3398	67.61 [57.01; 76.67]	698.1	98.40	<0.0001	0.0004	
Age < 5 years	22	7223	36,599	25.59 [21.03; 30.76]	2760.33	98.8	<0.0001	0.0781	NA
Period									0.1193
Pre-pandemic (2015/20)	19	5424	23,892	25.58 [20.94; 30.84]	1464.93	98.10	<0.0001	0.4885	
Lockdown (2020/21)	3	1422	11,913	17.34 [7.61; 34.84]	380.63	99.50	<0.0001	NA	
Post-lockdown (2021/22)	2	377	794	42.69 [24.30; 63.35]	57.54	98.30	<0.0001	NA	
Age < 10 years	4	7706	65,262	17.34 [11.89; 24.58]	1267.48	99.0	<0.0001	0.1347	NA
Period									0.0384
Pre-pandemic (2015/20)	4	5554	48,007	20.48 [13.90; 29.11]	1249.79	99.20	<0.0001	0.1212	
Lockdown (2020/21)	2	795	6989	7.39 [2.93; 17.40]	5.34	81.30	0.0208	NA	
Post-lockdown (2021/22)	1	1357	10,266	13.22 [12.58; 13.89]	0	NA	NA	NA	
Age < 15 years	20	5745	40,646	17.58 [12.82; 23.64]	4016.97	99.1	<0.0001	0.7428	NA
Period									<0.0001
Pre-pandemic (2015/20)	15	3963	31,279	19.46 [14.11; 26.23]	1939.82	98.70	<0.0001	0.0083	
Lockdown (2020/21)	7	304	5790	6.02 [2.38; 14.39]	168.99	96.40	<0.0001	NA	
Post-lockdown (2021/22)	6	1478	3577	31.37 [17.22; 50.12]	87.88	94.30	<0.0001	NA	
Age < 18 years	27	17,408	139,980	10.17 [7.29; 14.02]	5380.77	99.0	<0.0001	0.9850	NA
Period									<0.0001
Pre-pandemic (2015/20)	19	13,026	99,877	16.68 [13.29; 20.72]	3843.26	99.20	<0.0001	0.1671	
Lockdown (2020/21)	17	1695	23,124	2.53 [1.07; 5.85]	1230.46	98.70	<0.0001	0.1565	
Post-lockdown (2021/22)	7	2687	16,979	17.61 [14.30; 21.51]	48.9	87.70	<0.0001	NA	

Abbreviations: NA, not applicable.

3.6. Meta-Regression Model Analysis

Similar to subgroup analysis, the meta-regression model confirms that the prevalence of RSV during the lockdown period was lower than in the pre-pandemic period, as observed in the subgroup analysis (coefficient = 0.1526 [95% CI: 0.0856, 0.2195], $p < 0.0001$,

R^2 = 16.47%, with the lockdown period as a reference) (Table S9 in the Supplementary Material SA). On the other hand, this analysis indicates that RSV prevalence increased in the post-lockdown period, as also reported in the subgroup analysis (coefficient = 0.3005 [95% CI: 0.2059, 0.3951], p = 0.0001, R^2 = 16.47%, with the lockdown period as a reference) (Table S9 in the Supplementary Material SA).

Furthermore, the meta-regression analysis confirms that age and diagnostic technique significantly impact the prevalence of RSV. Age, especially among children under two years old, is a crucial factor influencing the magnitude of this effect. The bubble plot and meta-regression model further demonstrate a statistically significant negative relationship between RSV prevalence and age in months (coefficient = −0.0079 [−0.0105, −0.0054], $p < 0.0001$, R^2 = 22.29%) (Table S9 and Figure S3 in Supplementary Material SA). Additionally, the pooled prevalence of RSV remains unaffected by the risk of bias (coefficient −0.0554 [−0.1157, 0.0049], p = 0.0716, R^2 = 1.58%, with low risk as a reference), consistent with our subgroup analysis. Based on the AMSTAR 2 [17] criteria for assessing the quality of the body of evidence, the study meets all critical domains (items 2, 4, 7, 9, 11, 13, and 15), indicating a high-quality review (Table S10 in Supplementary Material SA).

3.7. Analysis of Intensive Care Unit Admissions and Mortality

We also analyzed the frequency of ICU admission and case fatality for children admitted with RSV infection. Most studies did not provide detailed information regarding these outcomes to allow assessment of the number of children with RSV infection who required admission to the ICU or who died. Although we had limited data on the prevalence of these outcomes, we conducted a meta-analysis using our dataset, which included 10 and 19 studies for ICU admission and RSV mortality, respectively. The analyses showed no significant differences in the prevalence of ICU admissions between the pre-pandemic and the lockdown periods (8.97% [95% CI: 2.60; 26.71] vs. 1.09% [95% CI: 0.15; 7.31], p = 0.07). The post-lockdown period was not included in these analyses due to the scarcity of data. Similarly, the comparison of mortality rates between the pre-pandemic, lockdown, and post-lockdown periods showed no significant difference (0.13% [95% CI: 0.01; 1.15], 3.57% [95% CI: 0.50; 21.42], and 0.0% [95% CI: 0.00; 100.00], respectively, p = 0.09) (Figures S4 and S5 in the Supplementary Material SA).

4. Discussion

This systematic review and meta-analysis aimed to assess the impact of NPIs during the COVID-19 pandemic on the prevalence of RSV in hospitalized children with LRTIs. Our comprehensive analysis of a large dataset provides valuable insights into the evolving prevalence of RSV across various pandemic phases and associated interventions.

Remarkably, our observations are similar to those of the meta-analysis conducted by Regassa et al. [26], which studied RSV prevalence in hospitalized children with LRTIs in Africa and reported a pooled prevalence of 23% [95% CI: 20.0; 25.0]. Similarly, a systematic review conducted by Pratt et al. [27] in hospitalized children with community-acquired pneumonia covering the pre-pandemic era showed an RSV pooled prevalence of 22.7% [95% CI: 20.9; 24.5], closely aligning with our results for the same period.

In most of our analyses, we confirmed a significant finding: a notable reduction in the prevalence of RSV during the lockdown period compared to the pre-pandemic period. This observation suggests that implementing NPIs such as social distancing, face masks, lockdowns, handwashing, and school closures to control the spread of SARS-CoV-2 had the unintended positive effect of effectively reducing the transmission of RSV.

The results of our study are consistent with previous research that suggests a decline in RSV-related hospitalizations during the COVID-19 pandemic [5,6]. For instance, Bardsley et al. [5], in an extensive study conducted in England focusing on children under five years of age and utilizing public health surveillance systems, reported an 80.8% decrease [95% CI: −80.9; −80.8] in admissions for RSV-attributable respiratory diseases from 2015–2019 to 2020–2021.

During the pandemic, similar trends were observed for other respiratory pathogens, demonstrating the efficacy of NPIs in controlling respiratory infections [28,29]. The substantial decline in positive tests for RSV and hospitalizations for LRTIs during the lockdown period can be attributed to reduced mobility and social interactions, which limited opportunities for RSV to spread among vulnerable children [7].

Interestingly, our review revealed a rebound effect as COVID-19 restrictions were eased. The prevalence of RSV in the post-lockdown period significantly increased compared to that in the lockdown period, and the proportion of RSV-associated hospitalizations surpassed that observed in the pre-pandemic period. This phenomenon was observed in various studies and countries, particularly during the summer of 2021 [5,9–12]. For example, Bardsley et al. [5] reported a staggering increase of 1258.3% [95% CI: 1178.3; 1345.8] in RSV cases in England. This increase could be attributed to the relaxation of NPIs, increased mobility, enhanced social interactions, and the subsequent spread of RSV [7].

Additionally, the population may have experienced a temporary decrease in immunity against RSV due to reduced exposure during the lockdown, a phenomenon known as "immunity debt", which could have rendered them more susceptible when the virus started circulating again [30–33]. Importantly, in Canada, Reicherz et al. [32] documented a significant reduction in prefusion RSV F IgG levels in women of childbearing age in 2021 compared to 2020 (148,858 $\pm$ 2.4 vs. 197,806 $\pm$ 2.2 AU/mL; p = 0.0232). Similarly, RSV-neutralizing titers in women decreased by 12-fold in 2021 compared with 2020 (10.3 $\pm$ 2.0 vs. 120.9 $\pm$ 2.9; $p < 0.0001$). Likewise, infants sampled in 2021 exhibited approximately 15-fold lower prefusion RSV F IgG levels (4258 $\pm$ 8.8 vs. 63,530 $\pm$ 4.4 AU/mL; $p < 0.0001$) and 3.4-fold lower RSV neutralizing titers (6.7 $\pm$ 1.8 vs. 22.8 $\pm$ 2.0; $p < 0.0001$) than infants sampled in 2020. In the Netherlands, den Hartog et al. [33] also found similar findings in a population of all ages (1–89 years). They observed that postfusion F RSV-specific IgG antibody concentrations declined from 2020 to 2021 ($p < 0.001$). These findings suggest a potential waning immunity that may have contributed to the global resurgence of RSV during interseasonal periods.

Another plausible explanation for the resurgence of RSV could be viral interference, whereby the high prevalence of SARS-CoV-2 during the lockdown period suppressed the circulation of RSV and other respiratory viruses [34]. As COVID-19 cases declined and NPIs were relaxed, RSV found a susceptible population and rapidly started spreading again.

Other proposed hypotheses regarding the change in RSV epidemiology during the pandemic include immune dysregulation due to SARS-CoV-2 infection, enhanced RSV virulence, and behavioral modifications in RSV testing practices among healthcare workers during the lockdown [35]. However, these proposed mechanisms require further in-depth study.

An important issue is to assess whether the low numbers of RSV infections documented during the lockdown period are the result of a reduction in RSV prevalence or a decrease in testing as a result of prioritizing SARS-CoV-2 detection, as healthcare resources were redirected towards managing COVID-19 during the pandemic [35]. Although a reduction in non-SARS-CoV-2 testing occurred in some countries, the available data suggest that the reduced proportion of RSV detections during the lockdown period was a result of reduced circulation of this virus. For example, a study carried out in Germany reported that during the period between December 2020 and March 2021, testing for four viruses (influenza A, influenza B, RSV, and SARS-CoV-2) was routinely used in an emergency department; during that period, 4915 tests were carried out and none were positive for influenza A, influenza B or RSV, despite the fact that the number of tests represented a five-fold increase compared to pre-pandemic figures [36]. Groves et al. analyzed the number of RSV tests carried out at sentinel laboratories in Canada during the 2020/2021 season and compared this with pre-pandemic testing; the weekly number of RSV tests during the 2020/2021 season (8890) was higher than the weekly average number of tests before the pandemic (6207), although the proportion of positive results was notably lower [37]. As shown in the abovementioned studies, even in countries where RSV testing was maintained or increased during the lockdown period, the number and proportion of RSV-positive tests were reduced.

RSV resurgence during the post-lockdown period highlights the importance of balancing the relaxation of NPIs and continuous viral monitoring with early interventions to manage the health impacts associated with outbreaks of respiratory infections. Future policies should consider a gradual easing of restrictions and ongoing monitoring to prevent sudden surges, particularly during respiratory virus seasons.

Our subgroup and meta-regression analyses identified age as a critical factor influencing the prevalence of RSV, with children under two years of age exhibiting the highest prevalence. This negative relationship between RSV prevalence and age underscores the vulnerability of young children to RSV infections. This observation is consistent with the understanding that young children, particularly infants and toddlers, are more susceptible to RSV due to their immature immune systems and narrower airways [38,39]. The higher prevalence of RSV among younger children highlights the need for targeted preventive strategies for this vulnerable population, particularly in post-lockdown scenarios.

The geographic emphasis of the study data suggests that the European region exhibited a higher RSV prevalence than other WHO regions. This regional difference in prevalence may be attributed to factors such as population density, climate, the level of healthcare surveillance, and adherence to NPIs [38,39]. Further studies are warranted to fully understand the regional variations in RSV prevalence during and after the pandemic.

In our review, determining which specific NPIs were more effective in mitigating the spread of SARS-CoV-2 and, consequently, RSV for each WHO region and age group was challenging. Factors such as pandemic severity, demographics, and local policies can influence their effectiveness, complicating the ability to carry out a comprehensive overview. In this context, Billard et al. [40] analyzed RSV surveillance data from 11 countries, focusing on the impact of nine NPIs from 2020 to 2021. They concluded that school closures, workplace closures, and stay-at-home measures were the most effective in reducing RSV spread. Similarly, Liu et al. [41] assessed the impact of NPIs on SARS-CoV-2 spread across over 130 countries using panel regression analysis and hierarchical cluster analysis, finding that school closures and restrictions on internal movement were consistently effective in all models. Additionally, Suryanarayanan et al. [42] used the Worldwide Non-pharmaceutical Interventions Tracker for COVID-19 (WNTRAC) to compile data on NPIs carried out in 261 countries and territories. WNTRAC showed that entertainment/cultural section closure (24.1%), confinement (15.0%), and school closures (13.9%) were the most common measures, aligning with the findings of Billard et al. [40] and Liu et al. [41]. These results suggest that school closures, workplace and entertainment closures, and confinement (stay-at-home) measures contributed to the reduction in the number of SARS-CoV-2 and RSV cases in most regions and countries.

Diagnostic techniques played a significant role in determining RSV prevalence estimates. Our study revealed that molecular assays, particularly qPCR, were associated with a higher RSV prevalence than immune assays. This finding is not unexpected, as molecular assays have higher sensitivity and specificity in detecting viral RNA [43,44]. Moreover, the use of nasopharyngeal secretions and sputum specimens was linked to a higher prevalence of RSV compared to other sample types, possibly due to the higher viral load present in the nasopharynx [43,44]. These findings are consistent with a systematic review by Regassa et al. [26] involving African children, which supports the importance of sample types in estimating RSV prevalence. This observation underscores the significance of diagnostic techniques in accurately assessing RSV prevalence and suggests that molecular assays could be more reliable for surveillance and clinical management.

In this review, the overall quality of evidence evaluated using the AMSTAR 2 [17] criteria was high. However, a critical limitation of this study was the presence of publication bias. This may be because most study were from the WHO regions of Europe and the Western Pacific, and only published studies were included. Arguably, this may limit the generalizability of our findings to more densely populated areas such as Southeast Asia, parts of the Americas, and sub-Saharan Africa. This concern could affect the validity of the conclusions drawn from the included studies, as the available data may not have been fully captured. Nevertheless, most of our meta-analyses focusing on the lockdown period (2020/21) showed no publication bias, suggesting more robust and reliable results within this subgroup. To comprehensively assess RSV prevalence during the COVID-19 pandemic, future studies should address publication bias and consider including unpublished studies with a more balanced distribution across WHO regions.

Additionally, significant heterogeneity was observed among the included studies in the meta-analysis, which can affect the precision of the pooled estimates. Subgroup and meta-regression analyses revealed several factors contributing to RSV prevalence heterogeneity, including age range, geographic region, sample type, and diagnostic techniques. Although random-effects models accounted for this heterogeneity, it is vital to interpret the findings cautiously. Our data demonstrated significant similarity to five systematic reviews and meta-analyses that focused on the prevalence of RSV in hospitalized children with LRTIs [26,27,45–47]. These studies also reported high heterogeneity, with I^2 values ranging from 90% to 99%, consistent with our findings.

Concerning the severity of RSV hospitalizations, our findings regarding ICU admissions and mortality are limited because few studies reported these outcomes in detail. No differences in ICU admissions and mortality were identified between the pre-pandemic, lockdown, and post-lockdown periods. Similarly, Nygaard et al. [48] analyzed a large dataset based on the Danish National Patient Registry. They reported that there were no differences in the use of mechanical ventilation in RSV-related hospitalizations in children (0–17 years) between the pre-pandemic and post-lockdown periods. It is important to underscore that our observations regarding this issue are limited. Thus, our results should be considered cautiously. This knowledge gap suggests that it is essential to study this topic in further detail in future research.

Finally, NPIs significantly reduced RSV cases; however, researchers have raised concerns about children's health in relation to NPIs, particularly regarding mental health issues such as anxiety and depression linked to quarantine and school closures. These measures also disrupt healthcare access, impacting the management of non-communicable diseases. Careful evaluation of the benefits and drawbacks of NPIs is crucial. In severe pandemics, where specific treatments or vaccines are unavailable, NPIs may be the only viable option. This necessitates policymakers to balance the benefits and risks of NPIs to ensure global well-being [49].

5. Conclusions

In summary, this systematic review and meta-analysis provides valuable evidence on the impact of NPIs during the COVID-19 pandemic on the prevalence of RSV in hospitalized children with LRTIs. The findings highlight the effectiveness of NPIs in reducing RSV transmission, particularly during periods of increased respiratory virus circulation. The resurgence of RSV following the relaxation of COVID-19 restrictions emphasizes the ongoing need for surveillance and public health interventions to mitigate the burden of respiratory infections in the post-pandemic era. These findings have important implications for public health policies and interventions to control RSV infections, particularly in vulnerable populations such as young children. Further research is required to investigate the long-term effects of NPIs on the transmission of RSV and to evaluate the potential benefits of vaccination and long-acting RSV monoclonal antibodies in conjunction with NPIs for reducing hospitalizations caused by RSV.

Supplementary Materials: The following supporting information can be downloaded at: https://www.mdpi.com/article/10.3390/v16030429/s1, Supplementary Material SA: Table S1: Preferred reporting items for systematic reviews and meta-analyses (PRISMA) 2020 checklist; Table S2: Search strategies from the PubMed, Scopus, and Web of Science databases; Table S3: Risk of bias tool by Hoy et al.; Table S4: Risk of bias assessment; Table S5: Countries of origin of the studies included in this systematic review and meta-analysis; Table S6: Characteristics of all included studies; Table S7:

Definitions of lower respiratory tract infection according to each study; Table S8: Sensitivity analysis of pooled respiratory syncytial virus (RSV) prevalence in children diagnosed with RSV-related lower respiratory tract infection using molecular assays; Table S9: Univariable meta-regression model for the prevalence of the respiratory syncytial virus in hospitalized children with lower respiratory tract infections; Table S10: The AMSTAR 2 (A Measurement Tool to Assess Systematic Reviews) instrument; Figure S1: Respiratory syncytial virus (RSV) prevalences according to RSV season; Figure S2: Funnel plot of sensitivity analysis of pooled respiratory syncytial virus (RSV) prevalence in children diagnosed with RSV-related lower respiratory tract infection by molecular assays: Figure S3: Bubble plot illustrating the association between average age in months and prevalence of respiratory syncytial virus; Figure S4. Prevalence of intensive care unit (ICU) admissions in hospitalized children with lower respiratory tract infections related to respiratory syncytial virus (RSV); Figure S5. Mortality among hospitalized children with lower respiratory tract infections related to respiratory syncytial virus (RSV), references of all 112 of the included studies are cited in the Tables S4, S6 and S7, Figures S1, S4 and S5; Supplementary Material SB: Table S1: Reasons for excluding full-text articles from the systematic review via databases, registers, and hand-searching.

Author Contributions: Conceptualization, D.E.N.; methodology, D.E.N. and J.J.L.-M.; software, J.J.L.-M., L.A.E.-M. and G.R.-A.; validation, D.E.N.; formal analysis, D.E.N. and J.J.L.-M.; investigation, D.E.N., L.A.E.-M., G.R.-A. and J.J.L.-M.; resources, D.E.N.; data curation, D.E.N., L.A.E.-M., G.R.-A. and J.J.L.-M.; writing—original draft preparation, J.J.L.-M., L.A.E.-M. and G.R.-A.; writing—review and editing, D.E.N. and J.J.L.-M.; visualization, D.E.N., L.A.E.-M. and G.R.-A.; supervision, D.E.N.; project administration, D.E.N.; funding acquisition, D.E.N. All authors have read and agreed to the published version of the manuscript.

Funding: This research received no external funding.

Institutional Review Board Statement: This is not applicable due to the nature of the systematic review, which involves the analysis of already published data without the direct involvement of humans.

Informed Consent Statement: This is not applicable due to the nature of the systematic review, which involves the analysis of already published data without the direct involvement of humans.

Data Availability Statement: All data are available in the main article and Supplementary Materials SA and SB.

Acknowledgments: José J. Leija-Martínez (CVU: 280319) expresses his gratitude to the Mexican National Council of Humanities, Science, and Technology (CONAHCYT) for the postdoctoral scholarship (Estancias Posdoctorales por México: 2022I1200/320/2022).

Conflicts of Interest: Daniel E. Noyola has participated as a member of the AbbVie, Sanofi Pasteur, and AstraZeneca speakers' bureau and served on the advisory board for Sanofi Pasteur, GSK, and Pfizer. The remaining authors declare no competing interests regarding this systematic review.

References

1. Li, Y.; Wang, X.; Blau, D.M.; Caballero, M.T.; Feikin, D.R.; Gill, C.J.; Madhi, S.A.; Omer, S.B.; Simões, E.A.F.; Campbell, H.; et al. Global, Regional, and National Disease Burden Estimates of Acute Lower Respiratory Infections Due to Respiratory Syncytial Virus in Children Younger Than 5 Years in 2019: A Systematic Analysis. *Lancet* **2022**, *399*, 2047–2064. [CrossRef] [PubMed]
2. Obando-Pacheco, P.; Justicia-Grande, A.J.; Rivero-Calle, I.; Rodríguez-Tenreiro, C.; Sly, P.; Ramilo, O.; Mejías, A.; Baraldi, E.; Papadopoulos, N.G.; Nair, H.; et al. Respiratory Syncytial Virus Seasonality: A Global Overview. *J. Infect. Dis.* **2018**, *217*, 1356–1364. [CrossRef] [PubMed]
3. Li, Y.; Reeves, R.M.; Wang, X.; Bassat, Q.; Brooks, W.A.; Cohen, C.; Moore, D.P.; Nunes, M.; Rath, B.; Campbell, H.; et al. Global Patterns in Monthly Activity of Influenza Virus, Respiratory Syncytial Virus, Parainfluenza Virus, and Metapneumovirus: A Systematic Analysis. *Lancet Glob. Health* **2019**, *7*, e1031–e1045. [CrossRef] [PubMed]
4. World Health Organization. Coronavirus Disease (COVID-19) Pandemic. Available online: https://www.who.int/emergencies/diseases/novel-coronavirus-2019?adgroupsurvey=%7Badgroupsurvey%7D&gclid=Cj0KCQiAgribBhDkARIsAASA5bsTI4_S1sMrB-dkdzOLNVFioBBozFxhlOz3V1XNgYAq7_OlBxqzluYaAgBGEALw_wcB (accessed on 11 November 2022).
5. Bardsley, M.; Morbey, R.A.; Hughes, H.E.; Beck, C.R.; Watson, C.H.; Zhao, H.; Ellis, J.; Smith, G.E.; Elliot, A.J. Epidemiology of Respiratory Syncytial Virus in Children Younger than 5 Years in England during the COVID-19 Pandemic, Measured by Laboratory, Clinical, and Syndromic Surveillance: A Retrospective Observational Study. *Lancet Infect. Dis.* **2023**, *23*, 56–66. [CrossRef] [PubMed]

6. Pruccoli, G.; Castagno, E.; Raffaldi, I.; Denina, M.; Barisone, E.; Baroero, L.; Timeus, F.; Rabbone, I.; Monzani, A.; Terragni, G.M.; et al. The Importance of RSV Epidemiological Surveillance: A Multicenter Observational Study of RSV Infection during the COVID-19 Pandemic. *Viruses* **2023**, *15*, 280. [CrossRef]
7. Ravkin, H.D.; Yom-Tov, E.; Nesher, L. The Effect of Nonpharmaceutical Interventions Implemented in Response to the COVID-19 Pandemic on Seasonal Respiratory Syncytial Virus: Analysis of Google Trends Data. *J. Med. Internet Res.* **2022**, *24*, e42781. [CrossRef]
8. Morawska, L.; Cao, J. Airborne Transmission of SARS-CoV-2: The World Should Face the Reality. *Environ. Int.* **2020**, *139*, 105730. [CrossRef]
9. Lavoie, P.M.; Reicherz, F.; Solimano, A.; Langley, J.M. Potential Resurgence of Respiratory Syncytial Virus in Canada. *CMAJ* **2021**, *193*, E1140–E1141. [CrossRef]
10. Foley, D.A.; Phuong, L.K.; Peplinski, J.; Lim, S.M.; Lee, W.H.; Farhat, A.; Minney-Smith, C.A.; Martin, A.C.; Mace, A.O.; Sikazwe, C.T.; et al. Examining the Interseasonal Resurgence of Respiratory Syncytial Virus in Western Australia. *Arch. Dis. Child.* **2022**, *107*, e7. [CrossRef] [PubMed]
11. McNab, S.; Ha Do, L.A.; Clifford, V.; Crawford, N.W.; Daley, A.; Mulholland, K.; Cheng, D.; South, M.; Waller, G.; Barr, I.; et al. Changing Epidemiology of Respiratory Syncytial Virus in Australia-Delayed Re-Emergence in Victoria Compared to Western Australia/New South Wales (WA/NSW) after Prolonged Lock-Down for Coronavirus Disease 2019 (COVID-19). *Clin. Infect Dis.* **2021**, *73*, 2365–2366. [CrossRef] [PubMed]
12. Curatola, A.; Graglia, B.; Ferretti, S.; Covino, M.; Pansini, V.; Eftimiadi, G.; Chiaretti, A.; Gatto, A. The Acute Bronchiolitis Rebound in Children after COVID-19 Restrictions: A Retrospective, Observational Analysis. *Acta Biomed.* **2023**, *94*, e2023031. [CrossRef]
13. Page, M.J.; McKenzie, J.E.; Bossuyt, P.M.; Boutron, I.; Hoffmann, T.C.; Mulrow, C.D.; Shamseer, L.; Tetzlaff, J.M.; Akl, E.A.; Brennan, S.E.; et al. The PRISMA 2020 Statement: An Updated Guideline for Reporting Systematic Reviews. *PLoS Med.* **2021**, *18*, e1003583. [CrossRef]
14. Noyola, D.E.; Esparza-Miranda, L.A.; Leija-Martínez, J.J.; Rivera-Alfaro, G. Impact of Non-Pharmaceutical Interventions During the COVID-19 Pandemic on the Prevalence and Severity of Respiratory Syncytial Virus Lower Respiratory Tract Infections in Pediatric Patients: A Systematic Review. *PROSPERO* **2022**, CRD42022376951. Available online: https://www.crd.york.ac.uk/prospero/display_record.php?ID=CRD42022376951 (accessed on 11 November 2022).
15. Higgins, J.P.T.; Thomas, J.; Chandler, J.; Cumpston, M.; Li, T.; Page, M.J.; Welch, V.A. (Eds.) Chapter 10: Analysing data and undertaking meta-analyses. In *Cochrane Handbook for Systematic Reviews of Interventions Version 6.3 (Updated February 2022)*; Cochrane: London, UK, 2022. Available online: www.training.cochrane.org/handbook (accessed on 11 November 2022).
16. Hoy, D.; Brooks, P.; Woolf, A.; Blyth, F.; March, L.; Bain, C.; Baker, P.; Smith, E.; Buchbinder, R. Assessing Risk of Bias in Prevalence Studies: Modification of an Existing Tool and Evidence of Interrater Agreement. *J. Clin. Epidemiol.* **2012**, *65*, 934–939. [CrossRef] [PubMed]
17. Shea, B.J.; Reeves, B.C.; Wells, G.; Thuku, M.; Hamel, C.; Moran, J.; Moher, D.; Tugwell, P.; Welch, V.; Kristjansson, E.; et al. AMSTAR 2: A Critical Appraisal Tool for Systematic Reviews That Include Randomised or Non-Randomised Studies of Healthcare Interventions, or Both. *BMJ* **2017**, *358*, j4008. [CrossRef]
18. Lin, L.; Chu, H. Meta-Analysis of Proportions Using Generalized Linear Mixed Models. *Epidemiology* **2020**, *31*, 713–717. [CrossRef]
19. Higgins, J.P.; Thompson, S.G.; Deeks, J.J.; Altman, D.G. Measuring Inconsistency in Meta-Analyses. *BMJ* **2003**, *327*, 557–560. [CrossRef]
20. Varela, F.H.; Sartor, I.T.S.; Polese-Bonatto, M.; Azevedo, T.R.; Kern, L.B.; Fazolo, T.; de David, C.N.; Zavaglia, G.O.; Fernandes, I.R.; Krauser, J.R.M.; et al. Rhinovirus as the Main Co-Circulating Virus During the COVID-19 Pandemic in Children. *J. Pediatr.* **2022**, *98*, 579–586. [CrossRef]
21. Lucion, M.F.; Del Valle Juárez, M.; Pejito, M.N.; Orqueda, A.S.; Romero Bollón, L.; Mistchenko, A.S.; Gentile, Á. Impact of COVID-19 on the Circulation of Respiratory Viruses in a Children's Hospital: An Expected Absence. *Arch. Argent. Pediatr.* **2022**, *120*, 99–105. [CrossRef]
22. Diesner-Treiber, S.C.; Voitl, P.; Voitl, J.J.M.; Langer, K.; Kuzio, U.; Riepl, A.; Patel, P.; Mühl-Riegler, A.; Mühl, B. Respiratory Infections in Children During a Covid-19 Pandemic Winter. *Front. Pediatr.* **2021**, *9*, 740785. [CrossRef] [PubMed]
23. Song, X.; Delaney, M.; Shah, R.K.; Campos, J.M.; Wessel, D.L.; DeBiasi, R.L. Common Seasonal Respiratory Viral Infections in Children Before and During the Coronavirus Disease 2019 (COVID-19) Pandemic. *Infect. Control Hosp. Epidemiol.* **2022**, *43*, 1454–1458. [CrossRef] [PubMed]
24. Ippolito, G.; La Vecchia, A.; Umbrello, G.; Di Pietro, G.; Bono, P.; Scalia Catenacci, S.; Pinzani, R.; Tagliabue, C.; Bosis, S.; Agostoni, C.; et al. Disappearance of Seasonal Respiratory Viruses in Children Under Two Years Old During COVID-19 Pandemic: A Monocentric Retrospective Study in Milan, Italy. *Front. Pediatr.* **2021**, *9*, 721005. [CrossRef] [PubMed]
25. Stera, G.; Pierantoni, L.; Masetti, R.; Leardini, D.; Biagi, C.; Buonsenso, D.; Pession, A.; Lanari, M. Impact of SARS-CoV-2 Pandemic on Bronchiolitis Hospitalizations: The Experience of an Italian Tertiary Center. *Children* **2021**, *8*, 556. [CrossRef] [PubMed]
26. Regassa, B.T.; Gebrewold, L.A.; Mekuria, W.T.; Kassa, N.A. Molecular Epidemiology of Respiratory Syncytial Virus in Children with Acute Respiratory Illnesses in Africa: A Systematic Review and Meta-Analysis. *J. Glob. Health* **2023**, *13*, 04001. [CrossRef] [PubMed]

27. Pratt, M.T.G.; Abdalla, T.; Richmond, P.C.; Moore, H.C.; Snelling, T.L.; Blyth, C.C.; Bhuiyan, M.U. Prevalence of Respiratory Viruses in Community-Acquired Pneumonia in Children: A Systematic Review and Meta-Analysis. *Lancet Child. Adolesc. Health* **2022**, *6*, 555–570. [CrossRef] [PubMed]
28. Vittucci, A.C.; Piccioni, L.; Coltella, L.; Ciarlitto, C.; Antilici, L.; Bozzola, E.; Midulla, F.; Palma, P.; Perno, C.F.; Villani, A. The Disappearance of Respiratory Viruses in Children During the COVID-19 Pandemic. *Int. J. Environ. Res. Public Health* **2021**, *18*, 9550. [CrossRef] [PubMed]
29. Rodgers, L.; Sheppard, M.; Smith, A.; Dietz, S.; Jayanthi, P.; Yuan, Y.; Bull, L.; Wotiz, S.; Schwarze, T.; Azondekon, R.; et al. Changes in Seasonal Respiratory Illnesses in the United States During the Coronavirus Disease 2019 (COVID-19) Pandemic. *Clin. Infect. Dis.* **2021**, *73*, S100–S107. [CrossRef] [PubMed]
30. Billard, M.N.; Bont, L.J. Quantifying the RSV Immunity Debt Following COVID-19: A Public Health Matter. *Lancet Infect. Dis.* **2023**, *23*, 3–5. [CrossRef]
31. Hatter, L.; Eathorne, A.; Hills, T.; Bruce, P.; Beasley, R. Respiratory Syncytial Virus: Paying the Immunity Debt with Interest. *Lancet Child. Adolesc. Health* **2021**, *5*, e44–e45. [CrossRef]
32. Reicherz, F.; Xu, R.Y.; Abu-Raya, B.; Majdoubi, A.; Michalski, C.; Golding, L.; Stojic, A.; Vineta, M.; Granoski, M.; Cieslak, Z.; et al. Waning Immunity Against Respiratory Syncytial Virus During the Coronavirus Disease 2019 Pandemic. *J. Infect. Dis.* **2022**, *226*, 2064–2068. [CrossRef]
33. den Hartog, G.; van Kasteren, P.B.; Schepp, R.M.; Teirlinck, A.C.; van der Klis, F.R.M.; van Binnendijk, R.S. Decline of RSV-Specific Antibodies During the COVID-19 Pandemic. *Lancet Infect. Dis.* **2023**, *23*, 23–25. [CrossRef]
34. Piret, J.; Boivin, G. Viral Interference between Respiratory Viruses. *Emerg. Infect. Dis.* **2022**, *28*, 273–281. [CrossRef] [PubMed]
35. Abu-Raya, B.; Viñeta Paramo, M.; Reicherz, F.; Lavoie, P.M. Why Has the Epidemiology of RSV Changed During the COVID-19 Pandemic? *EClinicalMedicine* **2023**, *61*, 102089. [CrossRef] [PubMed]
36. Stamm, P.; Sagoschen, I.; Weise, K.; Plachter, B.; Münzel, T.; Gori, T.; Vosseler, M. Influenza and RSV Incidence during COVID-19 Pandemic—An Observational Study from In-Hospital Point-of-Care Testing. *Med. Microbiol. Immunol.* **2021**, *210*, 277–282. [CrossRef] [PubMed]
37. Groves, H.E.; Piché-Renaud, P.P.; Peci, A.; Farrar, D.S.; Buckrell, S.; Bancej, C.; Sevenhuysen, C.; Campigotto, A.; Gubbay, J.B.; Morris, S.K. The Impact of the COVID-19 Pandemic on Influenza, Respiratory Syncytial Virus, and Other Seasonal Respiratory Virus Circulation in Canada: A Population-Based Study. *Lancet Reg. Health Am.* **2021**, *1*, 100015. [CrossRef] [PubMed]
38. Suleiman-Martos, N.; Caballero-Vázquez, A.; Gómez-Urquiza, J.L.; Albendín-García, L.; Romero-Béjar, J.L.; Cañadas-De la Fuente, G.A. Prevalence and Risk Factors of Respiratory Syncytial Virus in Children under 5 Years of Age in the WHO European Region: A Systematic Review and Meta-Analysis. *J. Pers. Med.* **2021**, *11*, 416. [CrossRef] [PubMed]
39. Johannesen, C.K.; van Wijhe, M.; Tong, S.; Fernández, L.V.; Heikkinen, T.; van Boven, M.; Wang, X.; Bøås, H.; Li, Y.; Campbell, H.; et al. Age-Specific Estimates of Respiratory Syncytial Virus-Associated Hospitalizations in 6 European Countries: A Time Series Analysis. *J. Infect. Dis.* **2022**, *226*, S29–S37. [CrossRef]
40. Billard, M.N.; van de Ven, P.M.; Baraldi, B.; Kragten-Tabatabaie, L.; Bont, L.J.; Wildenbeest, J.G. International Changes in Respiratory Syncytial Virus (RSV) Epidemiology during the COVID-19 Pandemic: Association with School Closures. *Influenza Other Respir. Viruses.* **2022**, *16*, 926–936. [CrossRef] [PubMed]
41. Liu, Y.; Morgenstern, C.; Kelly, J.; Lowe, R.; CMMID COVID-19 Working Group; Jit, M. The impact of non-pharmaceutical interventions on SARS-CoV-2 transmission across 130 countries and territories. *BMC Med.* **2021**, *19*, 40. [CrossRef]
42. Suryanarayanan, P.; Tsou, C.H.; Poddar, A.; Mahajan, D.; Dandala, B.; Madan, P.; Agrawal, A.; Wachira, C.; Samuel, O.M.; Bar-Shira, O.; et al. AI-Assisted Tracking of Worldwide Non-Pharmaceutical Interventions for COVID-19. *Sci. Data.* **2021**, *8*, 94. [CrossRef]
43. Onwuchekwa, C.; Atwell, J.; Moreo, L.M.; Menon, S.; Machado, B.; Siapka, M.; Agarwal, N.; Rubbrecht, M.; Aponte-Torres, Z.; Rozenbaum, M.; et al. Pediatric Respiratory Syncytial Virus Diagnostic Testing Performance: A Systematic Review and Meta-analysis. *J. Infect. Dis.* **2023**, *228*, 1516–1527. [CrossRef] [PubMed]
44. Bernstein, D.I.; Mejias, A.; Rath, B.; Woods, C.W.; Deeter, J.P. Summarizing Study Characteristics and Diagnostic Performance of Commercially Available Tests for Respiratory Syncytial Virus: A Scoping Literature Review in the COVID-19 Era. *J. Appl. Lab. Med.* **2023**, *8*, 353–371. [CrossRef]
45. Kenmoe, S.; Kengne-Nde, C.; Ebogo-Belobo, J.T.; Mbaga, D.S.; Fatawou Modiyinji, A.; Njouom, R. Systematic Review and Meta-analysis of the Prevalence of Common Respiratory Viruses in Children < 2 Years with Bronchiolitis in the Pre-COVID-19 Pandemic Era. *PLoS ONE* **2020**, *15*, e0242302. [CrossRef]
46. Umuhoza, T.; Bulimo, W.D.; Oyugi, J.; Musabyimana, J.P.; Kinengyere, A.A.; Mancuso, J.D. Prevalence of Human Respiratory Syncytial Virus, Parainfluenza and Adenoviruses in East Africa Community Partner States of Kenya, Tanzania, and Uganda: A Systematic Review and Meta-analysis (2007–2020). *PLoS ONE* **2021**, *16*, e0249992. [CrossRef]
47. Kenmoe, S.; Bigna, J.J.; Well, E.A.; Simo, F.B.N.; Penlap, V.B.; Vabret, A.; Njouom, R. Prevalence of Human Respiratory Syncytial Virus Infection in People with Acute Respiratory Tract Infections in Africa: A Systematic Review and Meta-analysis. *Influenza Other Respir. Viruses* **2018**, *12*, 793–803. [CrossRef]

48. Nygaard, U.; Hartling, U.B.; Nielsen, J.; Vestergaard, L.S.; Dungu, K.H.S.; Nielsen, J.S.A.; Sellmer, A.; Matthesen, A.T.; Kristensen, K.; Holm, M. Hospital admissions and need for mechanical ventilation in children with respiratory syncytial virus before and during the COVID-19 pandemic: A Danish nationwide cohort study. *Lancet Child. Adolesc. Health* **2023**, *7*, 171–179. [CrossRef] [PubMed]
49. ÓhAiseadha, C.; Quinn, G.A.; Connolly, R.; Wilson, A.; Connolly, M.; Soon, W.; Hynds, P. Unintended Consequences of COVID-19 Non-Pharmaceutical Interventions (NPIs) for Population Health and Health Inequalities. *Int. J. Environ. Res. Public Health* **2023**, *20*, 5223. [CrossRef] [PubMed]

viruses

MDPI

Article

Impact of SARS-CoV-2 Pandemic and Lockdown on the HRSV Circulation: Experience of Three Spoke Hospitals in Northern Italy

Francesca Parola [1,*], Adalberto Brach del Prever [1,*], Virginia Deut [2], Giulia Costagliola [3], Carla Guidi [1], Neftj Ragusa [2], Antonella Tuscano [3], Fabio Timeus [3] and Massimo Berger [2]

1 Pediatric and Neonatology Department, Ciriè Hospital, 10073 Ciriè, TO, Italy

2 Pediatric and Neonatology Department, Ivrea Hospital, 10015 Ivrea, TO, Italy; mberger@aslto4.piemonte.it (M.B.)

3 Pediatric and Neonatology Department, Chivasso Hospital, 10034 Chivasso, TO, Italy

* Correspondence: francesca.parola@aslto4.piemonte.it (F.P.); abrachdelprever@aslto4.piemonte.it (A.B.d.P.)

Abstract: The SARS-CoV-2 Pandemic affected the global epidemiology of respiratory infections, including Human Respiratory Syncytial Virus (HRSV), thanks to state governments' implementation of mitigation strategies, like the promotion of face masks and lockdowns. However, after the Pandemic, the dramatic resurge of these diseases was reported worldwide. Our retrospective study, involving three Spoke Pediatric Departments, includes all the infants under one year of age hospitalized for HRSV bronchiolitis in a period before the Pandemic period (2017–2020), during the SARS-CoV-2 Pandemic (2020–2021), and after the Pandemic (2021–2023). The primary aim was to analyze the temporal trend of HRSV in these three periods. Then, the clinical and epidemiological characteristics were analyzed to highlight the clinical differences in the affected patients, in the severity of the infections, and in the short-term outcomes. Ultimately, we analyzed the HRSV prevalence in the global bronchiolitis hospitalization over the reported periods. Overall, we included 237 patients. Before the Pandemic, the peak was recorded in January and February, while after the Pandemic, the peak was in November and December. A higher prevalence of HRSV was demonstrated after the Pandemic compared to the period before the Pandemic; overall, no difference in severity was reported. In conclusion, an increase in HRSV cases after the Pandemic has been demonstrated with an anticipated peak, while no differences were recorded in severity.

Keywords: Respiratory Syncytial Virus (HRSV); SARS-CoV-2 Pandemic; epidemiology; seasonality trend; HRSV prevalence

Citation: Parola, F.; Brach del Prever, A.; Deut, V.; Costagliola, G.; Guidi, C.; Ragusa, N.; Tuscano, A.; Timeus, F.; Berger, M. Impact of SARS-CoV-2 Pandemic and Lockdown on the HRSV Circulation: Experience of Three Spoke Hospitals in Northern Italy. *Viruses* **2024**, *16*, 230. https://doi.org/10.3390/10.3390/v16020230

Academic Editor: Donald Seto

Received: 29 December 2023
Revised: 26 January 2024
Accepted: 30 January 2024
Published: 1 February 2024

1. Introduction

Viral bronchiolitis is an acute respiratory illness that is the leading cause of hospitalization in young children; it represents the most common cause of acute respiratory failure in infants under one year of age in developing countries [1].

Many guidelines and consensus have been released to standardize the approach for viral bronchiolitis; the main point of discussion is regarding the definition, diagnosis, and treatment of bronchiolitis. Bronchiolitis has been diagnosed in children under one year of age, but the American guidelines use the diagnosis of bronchiolitis up to two years of age.

The diagnostic criteria of bronchiolitis, defined by an Italian Consensus referred to children under 12 months of age and included the following: an onset with rhinorrhea and/or upper respiratory tract infections; an episode of respiratory distress associated with crackles and/or wheezing, the use of accessory muscles or lower chest wall retractions, low O_2 saturation levels, high respiratory rate relative to age, skin color changes, nasal flaring, fever, and presentation during epidemic season [2].

Human Respiratory Syncytial Virus (HRSV) accounts for 60–80% of bronchiolitis presentations.

It has been estimated that HRSV infects more than 60% of all children during the first year of life and that HRSV infects nearly all children by the time they are two years old [3,4]. HRSV is the second cause of death worldwide after malaria and the first cause of death for respiratory illness [5,6].

The clinical presentation of HRSV infections is highly variable. It may be limited to upper respiratory tract symptoms, such as fever, rhinorrhea, and congestion; the severe presentation includes bronchiolitis and pneumonia. Bronchiolitis is generally self-limiting but can lead to hospitalization due to severe respiratory distress, acute respiratory failure, or difficulty in feeding. The main risk factors for severe disease are bronchopulmonary dysplasia, age younger than 12 months, a personal history of prematurity, male sex, immunodeficiency, formula feeding, and congenital heart disease.

The primary preventive measure used is prophylaxis with Palivizumab, a humanized monoclonal antibody against HRSV; Pavilizumab is used to defend against the severe manifestations of respiratory infections due to HRSV in high-risk patients, such as children born preterm or with bronchopulmonary dysplasia [7].

HRSV is characterized by a variable epidemiology, depending on geographic area. In Italy, the HRSV circulates from mild November until the end of March. It peaks in January/February; the total circulation duration is about four months [8].

Nevertheless, HRSV circulation in the past has been influenced by previous Pandemics; for instance, in 2009, influenza H1N1 delayed the HRSV peak. This variability of seasonality could be explained by the possible viral interference and the impact of preventive measures [9].

Concerning viral interference, a virus infection could be influenced by coinfection. Recently, during the Pandemic, the virus–virus interaction has been well defined; this interaction is influenced by the virus type, the infection timing, and the interplay between the response of the host to each virus interplay. A positive interaction is reported when co-infection results in an increased disease severity and pathogenesis, while a negative interaction is reported when an infection reduces or prevents the infection and replication of a second virus (e.g., influenza A virus and HRSV) [10].

During the SARS-CoV-2 Pandemic, various restrictive measures were adopted worldwide, like the imposition of social distancing measures, the closing of schools and commercial activities, strict hygiene behaviors, the use of face masks, and travel limitations. The massive effort to contain the spread of SARS-CoV-2 also affected the circulation of other respiratory pathogens, like influenza and HRSV, with a similar transmission route (contact, droplets, and aerosol transmission) [11].

Especially during the 2020–2021 season, a few cases of bronchiolitis were reported worldwide, leading some authors to speak about "a nearly absent disease" [12]. In the 2021–2022 season, a dramatic rebound of bronchiolitis was reported in the Northern and Southern hemispheres, due to HRSV infections [12]. This data in Italy was detected by a dedicated Surveillance Network system (RespivirNet), with a weekly report of respiratory viruses' circulation region-by-region [8].

The first report of decreased bronchiolitis was in Australia and New Zealand, where the containment of SARS-CoV-2 was excellent and quickly started to relax the SARS-CoV-2 preventive measures.

After, an unexpected unseasonal peak of bronchiolitis compared to that in the pre-Pandemic periods has been registered worldwide [13–15]. These data were confirmed by intercontinental reports; it shows an anticipation of the peak of bronchiolitis, due to HRSV. Also, in Italy, these data have been confirmed [16–19].

Our study aims to provide insights into the impact of the SARS-CoV-2 Pandemic on the epidemiology of HRSV infections in Spoke hospitals of our health district (ASLTO4). The secondary aim was to evaluate the differences in the clinical features of patients affected by

HRSV before and after the SARS-CoV-2 Pandemic and the global increase in bronchiolitis due to HRSV.

2. Materials and Methods

2.1. Study Design

A retrospective study was carried out in Spoke hospitals in the area nearing Turin, Piedmont Region, in the northwest of Italy. Italy is divided into health districts with a hub and spoke model for referral of patients. Our local health district (ASLTO4) includes 174 cities; the overall area is characterized by a great geographical variability, from high Alpine Mountains to urban areas. The total population of this area is 504,467 people, with 3015 inhabitants under one year of age in 2022 [20]. The local healthcare system ASLTO4 is organized into five districts, with great heterogeneity in demography, population density, and infrastructures. The General Emergency Department and Pediatric Unit are present in three Spoke hospitals of Ciriè, Chivasso, and Ivrea.

We included patients under one year of age referred to our Pediatric Departments for acute bronchiolitis due to HRSV that required hospitalization over different seasons. The test used to detect the HRSV was a rapid antigen test for the qualitative detection in nasopharyngeal swabs; the test used in our hospitals is the same since 2012 (BinaxNOW HRSV CARD—Abbot).

We divided the included patients into two groups: Group A hospitalized before the SARS-CoV-2 Pandemic (from 1 September 2017 to 31 March 2018; from 1 September 2018 to 31 March 2019; and from 1 September 2019 to 31 March 2020) and Group B hospitalized after the period of the Pandemic (from 1 September 2021 to 31 March 2022 and from 1 September 2022 to 31 March 2023) when the restrictive measures were relaxed.

In order to estimate the prevalence of HRSV, we reported the total number of cases of bronchiolitis hospitalized during the same periods and the few cases of bronchiolitis in the Pandemic period (from 1 September 2020 to 31 March 2021).

The period of data collection from September to March was established based on the HRSV epidemiology in Italy [8].

We collected data about demographic variables (sex, age, months of admission). Clinical and epidemiological data were recorded (age at onset, gestational age and birth weight, weight at admission, feeding, fever) as well as a personal history of chronic illnesses such as cardiopathy, neurological disease, and bronchopulmonary dysplasia. Relevant clinical variables were documented like laboratory (C-reactive protein in a performed blood sample and Pco2 in the Blood Gas Analysis) and microbiology (coinfection by other microbiological agents). C-reactive protein (CRP) was considered normal below 10 mg/dL. The coinfection was investigated based on clinical suspicion: we usually assessed antibodies for Epstein–Barr Virus or Mycoplasma pneumonia in blood samples. In Group B, all patients were tested for SARS-CoV-2 with an antigen rapid test on a nasopharyngeal swab for hospital admission. We reported radiograph results and short-term outcomes (length of hospital stay, complications, Hub hospital transfer). Treatment during hospitalization (low- or high-flow oxygen supplementation, nebulized therapy, steroids, antibiotics, and intravenous hydration) and discharged therapy data were recorded in all patients.

Firstly, the temporal trend of HRSV bronchiolitis after the SARS-COV-2 Pandemic (Group B) was compared to that of the previous period (Group A).

Then, the clinical and epidemiological characteristics were compared to assess if there were differences in the patients affected, in the severity of the infections, and the short-term outcomes.

Finally, we analyzed the HRSV prevalence in global bronchiolitis hospitalization in the same periods, both over-reported and during the Pandemic.

The study was conducted with full conformance to the principles of the Declaration of Helsinki. In accordance with the current legislation, this research is not among the types that requires a formal permission from an ethics committee. This is a secondary use of data for research purposes for which specific informed consent was requested ab initio from patients who would undertake a treatment process.

2.2. Statistical Analysis

Univariate analysis was performed with the Chi-square or Fisher's test for dichotomous variables, while the Kruskal–Wallis test for nonparametric measures was used for continuous variables [21]. The Kaplan–Meier statistics were used to define the probability of success [22]. The difference between groups was calculated with a log-rank test [23].

The univariate analysis was conducted using Vassar Stats (Statistical Computation Web Site), while the Kaplan–Meier statistics were performed using the NCSS software for Windows (https://www.ncss.com/; accessed on 4 December 2023). A p-value below 0.05 was defined as statistically significant.

3. Results

Overall, we hospitalized 468 bronchiolitis patients, 272 before the SARS-CoV-2 Pandemic (from 1 September 2017 to 31 March 2018; from 1 September 2018 to 31 March 2019; and from 1 September 2019 to 31 March 2020), 3 during the Pandemic (from 1 September 2020 to 31 March 2021), and 193 after the period of the Pandemic (from 1 September 2021 to 31 March 2022 and from 1 September 2022 to 31 March 2023).

In 237 patients, the HRSV was detected on the nasal swab, and for this reason, they were enrolled in the study: 109 in the Group pre-pandemic (named Group A) and 128 in the Group post-pandemic (named Group B). No HRSV bronchiolitis patient was admitted during the SARS-CoV-2 pandemic.

3.1. Seasonality

Hospitalization for HRSV bronchiolitis followed a distinct seasonal trend in the two groups, as shown in Figure 1.

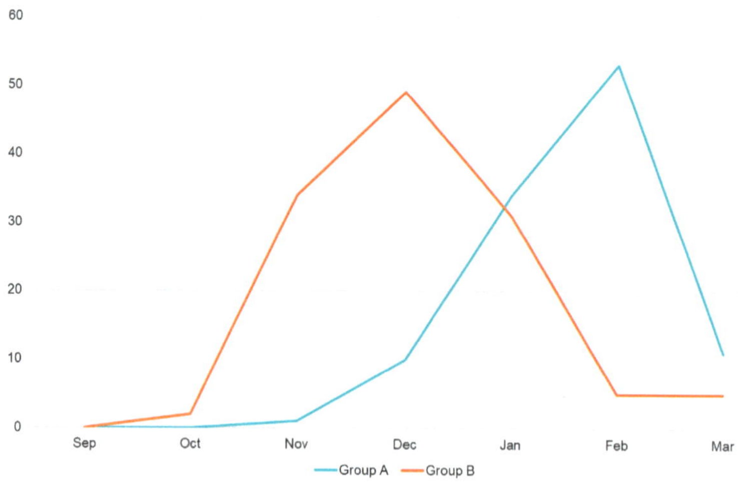

Figure 1. Seasonal trend of HRSV bronchiolitis.

In Group A (patients admitted in the period before the SARS-CoV-2 pandemic), the peaks in admissions occurred between November and March, usually lasting 2–4 months. Cases increased significantly in December and peaked in January and February, with only a few cases reported in March.

On the other hand, in Group B (patients admitted after the SARS-CoV-2 pandemic), an anticipated peak was reported. Indeed, bronchiolitis started slowly in October, peaked in November and December, and slowly decreased during February and March.

The distribution of hospitalization month-by-month is summarized in Table 1.

Table 1. Distribution of the seasonality of infections in the two groups.

	Group A N = 109	Group B N = 128	*p*-Value
September	0 (0%)	0 (0%)	
October	0 (0%)	2 (2%)	
November	1 (1%)	36 (28%)	
December	10 (9%)	49 (38%)	<0.0001
January	34 (31%)	31 (24%)	
February	53 (49%)	5 (4%)	
March	11 (10%)	5 (4%)	

3.2. Clinical and Epidemiological Characteristics of Hospitalized Patients

The demography and clinical characteristics of patients hospitalized in all periods that were considered (pre-pandemic, during pandemic, and after pandemic) are provided in Table 2.

Table 2. Demographic and clinical characteristics of patients.

		Group A N = 109	Group B N = 128	*p*-Value
Gender	Male	56 (51%)	78 (61%)	0.14
	Female	53 (49%)	50 (39%)	
Median Age (months)		2 (IQR 1–3)	2 (IQR 1–4)	0.85
Prematurity	Yes	12 (11%)	15 (12%)	0.36
	No	95 (87%)	106 (83%)	
	Unknown	2 (2%)	7 (5%)	
Months at diagnosis	<1	37 (34%)	62 (48%)	0.0012
	1–3	52 (48%)	30 (23%)	
	4–6	14 (13%)	22 (17%)	
	6–12	6 (5%)	14 (11%)	
Weight at diagnosis	<4000 g	21 (19%)	22 (17%)	0.21
		31 (28%)	34 (26%)	
	>5000 g	57 (52%)	69 54%)	
Feeding	Breast-feed	63 (58%)	84 (66%)	0.059
	Artificial	37 (34%)	26 (20%)	
	Weaned baby	7 (6%)	13 (10%)	
Comorbidity	Yes	6 (5%)	10 (8%)	0.6
	No	102 (93%)	118 (92%)	
Coinfection	Yes	1 (1%)	3 (4%)	0.62
	No	108 (99%)	125 (96%)	

Table 2. *Cont.*

		Group A N = 109	Group B N = 128	*p*-Value
Fever	Yes	45 (41%)	31 (24%)	0.0054
	No	64 (59%)	97 (76%)	
Profilaxed with Palivizumab		0 (0%)	1 (0,78%)	1

In Group A, males were more represented than females (51% vs. 49%); gestational age was reported in 98% of patients. Preterm infants accounted for about 11%; half of the patients had a gestational age between 30 and 34 weeks, and the remaining patients had a gestational age between 35 and 36 weeks. In this group (patients admitted before SARS-SARS-CoV-2 pandemic), no patient had a gestational age under 30 weeks. Two patients had bronchopulmonary dysplasia, two infants had chronic heart conditions, one had a neurological disease like epilepsy, and no infant with immunodeficiency was recorded. At the time of diagnosis, most patients (48%) had an age between one and three months while 34% of them were under one month of age; furthermore, 13% had an age between four and six months and 5% of them were over six months of age. Breastfeeding was reported in 58% of all patients, while formula feeding was reported in 34% of patients of the Group A. Furthermore, 6% of infants had already been weaned. In one patient, during the hospitalization, a coinfection of Epstein–Barr Virus (EBV) was detected.

As specified before, no HRSV bronchiolitis was found during the SARS-CoV-2 pandemic in our hospitals.

In Group B (patients admitted after the SARS-CoV-2 pandemic), more males than females (61% vs. 39%) were represented, and the gestational age was reported in 95% of patients. Among the patients for whom the gestational age was available, preterm infants were calculated to be 12% of all patients; most of these patients (67%) had a gestational age between 35 and 36 weeks, while 27% of them had a gestational age between 30 and 34 weeks, and only one infant had a gestational age of 25 weeks. Two patients had bronchopulmonary dysplasia, and one infant had a neurological disease with hypotonia. No infants affected by immunodeficiency or chronic heart conditions were recorded. When hospitalized, most patients (48%) were newborns under one month of age, while 23% had an age between one and three months, and 17% were between four and six months. Overall, 11% of infants were over six months old of age. Breastfeeding was reported in 66% of all patients, while formula feeding was reported as feeding modality in 20% of all patients. Furthermore, 10% of infants had already been weaned. Concerning the coinfections, we reported three cases in whom another pathogen was found. In one case, we also found an EBV first infection; in the second patient, a Mycoplasma pneumoniae co-infection occurred, and in the last patient, the nasal swab tested positive for both HRSV and SARS-CoV-2.

In the description of the demographic features of patients involved in the study, we found a remarkable difference in the age of the infants; in Group B (post-pandemic), there were more newborns. No statistical differences in sex, gestational age, feeding, and comorbidity were recorded.

Concerning the prophylaxis with Palivizumab, only one patient in Group B was hospitalized for bronchiolitis due to HRSV; he was born at 31 weeks of gestational age and needed Continuous Positive Airway Pressure (CPAP) at birth. He also presented pneumothorax in the hours after birth. During the hospitalization for bronchiolitis, he did not need oxygen supplementation and was admitted to the hospital for three days.

Laboratory, X-ray findings, and treatment during hospitalization are provided in Table 3.

Table 3. Laboratory, X-ray findings, and treatment during hospitalization.

		Group A **N = 109**	**Group B** **N = 128**	***p*-Value**
		1.98 (0–63.5)	1.19 (0–60.3)	0.039
Chest X-rays (number)	Yes	23 (21%)	19 (15%)	0.23
	No	86 (79%)	109 (85%)	
Chest X-rays (pathologic results)		13 (12%)	10 (8%)	0.27
Higher level of pCO2 (mmHg)		46 (28.2–64.1)	48 (31.7–74)	1
Low-flow Oxygen	Yes	64 (59%)	79 (62%)	0.69
	No	45 (41%)	49 (38%)	
Low-flow Oxygen days		3 (0–8)	3 (0–9)	0.76
High-flow Oxygen	Yes	29 (27%)	33 (26%)	1
	No	80 (73%)	94 (73%)	
High-flow Oxygen days		4 (1–8)	5 (1–11)	0.38
Fraction of inspired Oxygen %		35 (21–50)	34 (25–65)	0.56
Aerosolized drugs	Yes	96 (88%)	122 (95%)	0.054
	No	13 (12%)	6 (5%)	
Antibiotics	Yes	42 (38%)	24 (19%)	0.00082
	No	67 (61%)	104 (81%)	
Steroids	Yes	13 (12%)	17 (13%)	0.84
	No	96 (88%)	111 (87%)	
Hydration IV	Yes	32 (29%)	23 (18%)	0.045
	No	77 (71%)	105 (82%)	

Notes: CRP: C-reactive protein; pCO_2: partial pressure of carbon dioxide; and IV: intravenous.

There were no significant differences between the two groups in X-ray findings, while there was a remarkable difference in CRP values.

Concerning the treatment, some differences were remarkable. Antibiotics and IV Hydration were less used in Group B; in Group A, 38% received an antibiotic therapy compared to the 19% of patients in Group B. Similarly, in Group B, only 18% of patients were treated with IV hydration (vs. 29% in Group A).

There were no statistically significant differences in the type of respiratory support and the length of respiratory support in days.

Finally, concerning the short-term outcome, the complications, the length of hospital stay, and the need to transfer to a Hub hospital were similar in the two groups. These data are provided in Table 4. Figure 2 reports the probability of treatment success (discharge to home without transfer to a Hub hospital with PICU or NICU): 96% of patients in Group A and 94% in Group B.

Table 4. Short-term outcome.

		Group A N = 109	Group B N = 128	*p*-Value
Complications	Yes	12 (11%)	12 (9%)	0.82
	No	97 (89%)	116 (91%)	
Hospitalization (days)		6 (2–33)	5 (1–13)	0.051
Transfer to a Hub hospital	Yes	4 (4%)	7 (5%)	0.55
	No	105 (96%)	121 (94%)	

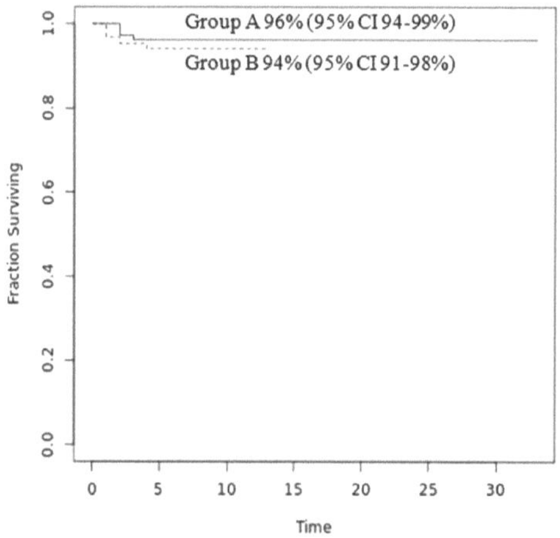

Figure 2. Probability of treatment success (discharge to home). Notes: gray line: Group A and light blue line: Group B.

3.3. HRSV Prevalence

In the period before the SARS-CoV-2 Pandemic (from 1 September 2017 to 31 March 2018; from 1 September 2018 to 31 March 2019; and from 1 September 2019 to 31 March 2020) in the three ASLTO4 Pediatric Departments, 272 patients were hospitalized with the diagnosis of bronchiolitis, and in 109 (40%) of them, HRSV (Group A) was detected on the nasal swab.

During the SARS-CoV-2 Pandemic (from 1 September 2020 to 31 March 2021), when restrictive measures were adopted, 3 patients were hospitalized for bronchiolitis, and HRSV was not detected.

After the SARS-CoV-2 pandemic (from 1 September 2021 to 31 March 2022 and from 1 September 2022 to 31 March 2023), the total number of hospitalizations for bronchiolitis were 193. In 128 cases (66%), the detected agent was HRSV (Group B).

Therefore, we can demonstrate a significant increase in the prevalence of HRSV after the SARS-CoV-2 Pandemic; moreover, we noticed a significant drop in the cases of bronchiolitis, but, above all, we found the complete absence of HRSV infection in the Pandemic period.

All these data are summarized in Table 5 and Figures 3 and 4.

Table 5. Cases of HRSV infections in patients hospitalized for bronchiolitis.

	2017–2020	2021–2023	*p*-Value
HRSV positive	109 (40%)	128 (66%)	<0.0001
HRSV negative	163 (60%)	65 (34%)	

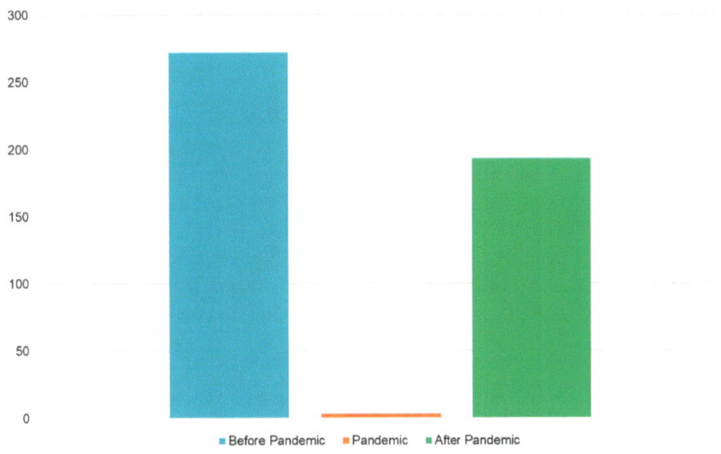

Figure 3. Hospitalization for bronchiolitis before, during, and after SARS-CoV-2 Pandemic.

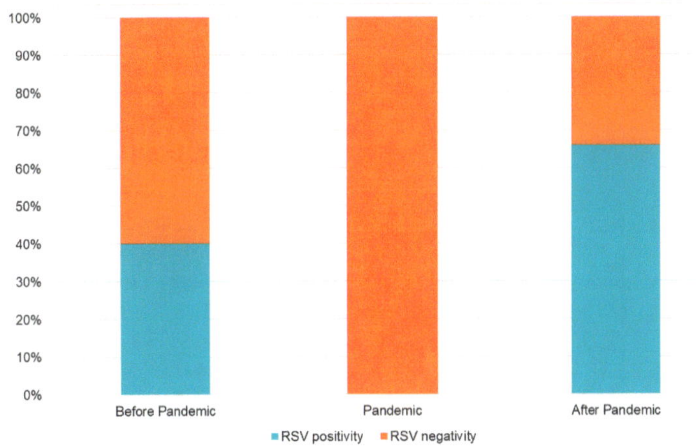

Figure 4. Prevalence of HRSV detected in bronchiolitis hospitalized before, during, and after the SARS-CoV-2 Pandemic.

4. Discussion

The Pandemic shows the importance of respiratory viruses' circulation surveillance. Respiratory infections are very common in infants and their severity varies based on the host features (for instance, the presence of prematurity or cardio-pulmonary broncho dysplasia). Above all causes of respiratory infection, viral bronchiolitis is the most frequent lower respiratory tract infection and the leading cause of hospitalization in children less than twelve months of age. HRSV is the virus most involved in severe bronchiolitis, and its prevalence shows territorial differences.

Our study analyzes the characteristics of infants hospitalized for acute HRSV bronchiolitis in the three Pediatric Departments of ASLTO4 before, during, and after the SARS-CoV-2 Pandemic.

Our data highlight that after the SARS-CoV-2 Pandemic, the HRSV epidemic started earlier than usual. We showed a peak in November–December in Group B (post-pandemic), while in Group A (pre-pandemic), the peak was reached in January–February. This confirms new epidemiological trends of HRSV infection as reported worldwide [18,24–27].

The surveillance of the seasonality of HRSV is very important for the improvement and adaptation of the prevention measures. The logistics and timing are very important to optimize its prevention results; if the prophylaxis begins months before the HRSV season, the protection could wane before the end of the epidemic, leaving infants susceptible to HRSV. Similarly, if the HRSV season starts earlier than the prophylaxis, high-risk infants remain vulnerable. Considering the new epidemiological trends, in Piedmont, prophylaxis with Palivizumab in high-risk patients begins in October, starting from 2022, providing the first dose before the onset of the circulation of HRSV. This highlights the importance of improving and updating the system of local surveillance that started in Italy in 2019–2020 as the Influenza Surveillance Network (Respirvirnet) system.

In accordance with the current literature, no differences in the severity of HRSV infection have been demonstrated in our study; other authors have reported no differences in X-ray findings, short-term outcomes (like complications, the length of hospitalization, the type of respiratory support, the length of oxygen supplementation, and the need of transfer to a Hub hospital) [28–30]. The only remarkable difference in our study is in the results of the CRP dosage; in Group A (pre-pandemic), its average value was higher than that in Group B (post-pandemic).

Other studies, similar to this one, were conducted in Italy. It is very interesting to note that studies conducted in Sicily, a region in the South of Italy, have a similar trend of HRSV infection. The authors did not similarly report a variation in the disease severity [31,32]. Piedmont (in the northwest of Italy) and Sicily (an island in the south of Italy) have very different climate and air characteristics. The similar results in HRSV trends suggest that even in parts of Italy where there is different weather due to the different latitudes, the HRSV epidemiological trend, the severity of the disease, and its prevalence are similar.

Interestingly, we showed remarkable differences in treatment; fewer antibiotics were used in Group B. This suggests better adherence to the guidelines, making an important effort to reduce the development of antibiotic resistance. On the other hand, a higher CRP value was found in Group A; this finding may explain the higher use of antibiotics in this group of patients: clinicians had more suspicion of bacterial complications, so they prescribed antibiotics. Lastly, by reducing the rate of IV hydration over the years, the patients were encouraged to maintain oral feeding with a higher number of fractionated meals, as suggested by the national guidelines [4].

Looking at the HRSV prevalence, during the SARS-CoV-2 pandemic, we observed a drop in bronchiolitis hospitalizations. We reported three patients that were hospitalized for bronchiolitis, but no case of HRSV infection was found. These data are in agreement with other studies and the epidemiological surveillance of viruses that show a global reduction in respiratory infections that share the transmission path with SARS-CoV-2, like influenza and HRSV [26,33]. Given that the transmission of HRSV occurs through droplets, the containment measures of the SARS-CoV-2 Pandemic (like the use of face masks, social distancing, smart working, and the closure of schools) have led to the reduction in HRSV transmission [11,34,35].

Moreover, there is a strong correlation between environmental conditions, for example, weather and air pollution, and the incidence of HRSV. A study shows a correlation between HRSV transmission, benzene levels, and humidity, while there is an inverse correlation with temperature [36]. Given that a significant reduction in air pollutants such as benzene was recorded during the Pandemic, it can be hypothesized that the pollution reduction helped to decrease the circulation of HRSV. It is very hard to quantify the contribution, but we can suggest that it is not comparable with restrictive measures. We aim to explore such hypotheses in future studies.

The prevalence of HRSV in our study group increased significantly after the SARS-CoV-2 Pandemic, confirming previously published data; we reported an HRSV surge when the prevention measures were relaxed, and the social interactions increased. This finding confirms the main role of social distance measures in the containment of HRSV circulation.

Furthermore, the "immunity debt" played a role in the increased circulation of HRSV: during the Pandemic period, the cohort of HRSV-naïve patients expanded. It happened for two reasons: (1) due to the presence of children who have never had HRSV infections and (2) due to the reduction in immunity duration, which decreased during the time and without re-exposure to HRSV. This is confirmed by studies that show an increased number of older infants affected by HRSV [36].

In our study, there is a remarkable difference in newborns hospitalized for HRSV bronchiolitis comparing pre- and post-pandemic, with 48% of newborns in Group B vs. 34% of newborns in Group A. This result is due to the reduced exposure to respiratory viruses not only in children but also in pregnant women; above all, it is known that the infection in the third trimester of pregnancy could protect newborns against HRSV infections through antibodies contained in breast milk and those transferred transplacentally [37–39].

Our study has some limitations. Our analysis is conducted in Hub hospitals with a small sample size, due to the characteristics of our health district as detailed below: First of all, a small percentage of children aged less than one year live in our district. Furthermore, the study is based on a retrospective data collection of hospital records; as a consequence, this study may be subject to information bias due to the lack of data or incomplete hospital records. In addition, we only reported data about hospitalized patients with bronchiolitis, while patients who visited the Emergency Department and were discharged or treated by general pediatric practitioners were not included. Consequently, the global prevalence and the incidence of HRSV may be underestimated.

The main strength of our study is that it was performed in three Spoke hospitals that are a part of the same health district (ASL); therefore, clinicians had the same devices for the treatment of respiratory failure and the same tests to detect HRSV. The test used was an antigen test that was less sensitive and specific than molecular methods, but the same test was used in the three hospitals and in the different analyzed periods. Furthermore, clinicians used the same protocols of treatment and had the same criteria of transfer to Hub Hospital because of the lack of Pediatric Intensive Care Unit in our health district.

5. Conclusions

The SARS-CoV-2 Pandemic has changed the epidemiological trend of HRSV infections in our territory. In detail, the bronchiolitis season started earlier than usual after the Pandemic; this is reported in our study in accordance with the data records from other countries. The unusual resurgence of HRSV infection was not associated with an increased severity of the illness in our study group. In addition, we reported an increase in the prevalence of HRSV bronchiolitis hospitalized after the Pandemic, with a high proportion of newborns possibly due to the "immunity debt" and the lower exposure in pregnant women.

The surveillance of the circulation of the respiratory virus is necessary to adapt the preventive measures and the hospital activity organization to the seasonal changes; indeed, the analysis of changes in seasonality allows high-risk patients to receive the optimal level of prevention with the correct prophylaxis while hospitals reorganize their activities. It may also imply transferring patients from the Hub to the Spoke hospitals, leading to a remarkable reduction in costs. Lastly, a lesson learned during the Pandemic period was that the simple preventive measures should not be forgotten, because they can markedly reduce HRSV circulation; this finding underlines the importance of the strict hygiene behaviors and the utilization of face masks by healthcare workers, who predominantly deal with high-risk patients.

Author Contributions: F.P., A.B.d.P., V.D., G.C., C.G., N.R., A.T., F.T. and M.B. contributed equally to all sections of the paper, including contribution to data curation, to the analysis of the results and to the writing of the manuscript. F.P. and M.B. contributed to the draft. All authors have read and agreed to the published version of the manuscript.

Funding: This research received no external funding.

Institutional Review Board Statement: The study was conducted in accordance with the Declaration of Helsinki. In accordance with the current legislation, this research is not among the types that requires a formal permission from an ethics committee.

Informed Consent Statement: Not applicable.

Data Availability Statement: The collected data are available from the corresponding authors.

Conflicts of Interest: The authors declare no conflicts of interest.

References

1. Meissner, H.C. Viral Bronchiolitis in Children. *N. Engl. J. Med.* **2016**, *374*, 62–72. [CrossRef]
2. Baraldi, E.; Lanari, M.; Manzoni, P.; Rossi, G.A.; Vandini, S.; Rimini, A.; Romagnoli, C.; Colonna, P.; Biondi, A.; Biban, P.; et al. Inter-society consensus document on treatment and prevention of bronchiolitis in newborns and infants. *Ital. J. Pediatr.* **2014**, *40*, 65. [CrossRef] [PubMed]
3. Perez, A.; Lively, J.Y.; Curns, A.; Weinberg, G.A.; Halasa, N.B.; Staat, M.A.; Szilagyi, P.G.; Stewart, L.S.; McNeal, M.M.; Clopper, B.; et al. Respiratory Virus Surveillance Among Children with Acute Respiratory Illnesses—New Vaccine Surveillance Network, United States, 2016–2021. *MMWR Morb. Mortal. Wkly. Rep.* **2022**, *71*, 1253–1259. [CrossRef]
4. Manti, S.; Staiano, A.; Orfeo, L.; Midulla, F.; Marseglia, G.L.; Ghizzi, C.; Zampogna, S.; Carnielli, V.P.; Favilli, S.; Ruggieri, M.; et al. UPDATE—2022 Italian guidelines on the management of bronchiolitis in infants. *Ital. J. Pediatr.* **2023**, *49*, 19. [CrossRef]
5. Cutrera, R.; Wolfler, A.; Picone, S.; Rossi, G.A.; Gualberti, G.; Merolla, R.; Del Vecchio, A.; Villani, A.; Midulla, F.; Dotta, A. Impact of the 2014 American Academy of Pediatrics recommendation and of the resulting limited financial coverage by the Italian Medicines Agency for palivizumab prophylaxis on the RSV-associated hospitalizations in preterm infants during the 2016–2017 epidemic season: A systematic review of seven Italian reports. *Ital. J. Pediatr.* **2019**, *45*, 139. [CrossRef]
6. Mazur, N.I.; Martinón-Torres, F.; Baraldi, E.; Fauroux, B.; Greenough, A.; Heikkinen, T.; Manzoni, P.; Mejias, A.; Nair, H.; Papadopoulos, N.G.; et al. Lower respiratory tract infection caused by respiratory syncytial virus: Current management and new therapeutics. *Lancet Respir. Med.* **2015**, *3*, 888–900. [CrossRef] [PubMed]
7. Viguria, N.; Navascués, A.; Juanbeltz, R.; Echeverría, A.; Ezpeleta, C.; Castilla, J. Effectiveness of palivizumab in preventing respiratory syncytial virus infection in high-risk children. *Hum. Vaccines Immunother.* **2021**, *17*, 1867–1872. [CrossRef] [PubMed]
8. Azzari, C.; Baraldi, E.; Bonanni, P.; Bozzola, E.; Coscia, A.; Lanari, M.; Manzoni, P.; Mazzone, T.; Sandri, F.; Lisi, G.C.; et al. Epidemiology and prevention of respiratory syncytial virus infections in children in Italy. *Ital. J. Pediatr.* **2021**, *47*, 198. [CrossRef]
9. Li, Y.; Wang, X.; Msosa, T.; de Wit, F.; Murdock, J.; Nair, H. The impact of the 2009 influenza pandemic on the seasonality of human respiratory syncytial virus: A systematic analysis. *Influenza Other Respir. Viruses* **2021**, *15*, 804–812. [CrossRef]
10. Piret, J.; Boivin, G. Viral Interference between Respiratory Viruses. *Emerg. Infect. Dis.* **2022**, *28*, 273–281. [CrossRef]
11. Nascimento, M.S.; Baggio, D.M.; Fascina, L.P.; Prado, C.D. Impact of social isolation due to COVID-19 on the seasonality of pediatric respiratory diseases. *PLoS ONE* **2020**, *15*, e0243694. [CrossRef]
12. Van Brusselen, D.; De Troeyer, K.; Ter Haar, E.; Vander Auwera, A.; Poschet, K.; Van Nuijs, S.; Bael, A.; Stobbelaar, K.; Verhulst, S.; Van Herendael, B.; et al. Bronchiolitis in COVID-19 times: A nearly absent disease? *Eur. J. Pediatr.* **2021**, *180*, 1969–1973. [CrossRef]
13. Foley, D.A.; Yeoh, D.K.; Minney-Smith, C.A.; Martin, A.C.; Mace, A.O.; Sikazwe, C.T.; Le, H.; Levy, A.; Moore, H.C.; Blyth, C.C. The Interseasonal Resurgence of Respiratory Syncytial Virus in Australian Children Following the Reduction of Coronavirus Disease 2019–Related Public Health Measures. *Clin. Infect. Dis.* **2021**, *73*, e2829–e2830. [CrossRef]
14. Saravanos, G.L.; Hu, N.; Homaira, N.; Muscatello, D.J.; Jaffe, A.; Bartlett, A.W.; Wood, N.J.; Rawlinson, W.; Kesson, A.; Lingam, R.; et al. RSV Epidemiology in Australia Before and During COVID-19. *Pediatrics* **2022**, *149*, e2021053537. [CrossRef]
15. Public Health Surveillance Inforamtion for New Zealands Public Health Action. Laboratory-Based Virology Weekly Report, 2004–2019. 2021. Available online: https://surv.esr.cri.nz/virology/virology_weekly_report.php (accessed on 28 December 2023).
16. Camporesi, A.; Morello, R.; Ferro, V.; Pierantoni, L.; Rocca, A.; Lanari, M.; Trobia, G.L.; Sciacca, T.; Bellinvia, A.G.; De Ferrari, A.; et al. Epidemiology, Microbiology and Severity of Bronchiolitis in the First Post-Lockdown Cold Season in Three Different Geographical Areas in Italy: A Prospective, Observational Study. *Children* **2022**, *9*, 491. [CrossRef] [PubMed]
17. Tulyapronchote, R.; Selhorst, J.B.; Malkoff, M.D.; Gomez, C.R. Delayed sequelae of vertebral artery dissection and occult cervical fractures. *Neurology* **1994**, *44*, 1397. [CrossRef] [PubMed]
18. Agha, R.; Avner, J.R. Delayed Seasonal RSV Surge Observed During the COVID-19 Pandemic. *Pediatrics* **2021**, *148*, e2021052089. [CrossRef]
19. Guitart, C.; Bobillo-Perez, S.; Alejandre, C.; Armero, G.; Launes, C.; Cambra, F.J.; Balaguer, M.; Jordan, I.; Pagarolas, A.A.; Vila, J.; et al. Bronchiolitis, epidemiological changes during the SARS-CoV-2 pandemic. *BMC Infect. Dis.* **2022**, *22*, 84. [CrossRef] [PubMed]
20. Available online: http://www.ruparpiemonte.it/infostat/risultati.jsp (accessed on 28 December 2023).
21. Kaplan, E.L.; Meier, P. Nonparametric Estimation from Incomplete Observations. *J. Am. Stat. Assoc.* **1958**, *53*, 457–481. [CrossRef]
22. Mantel, N. Evaluation of survival data and two new rank order statistics arising in its consideration. *Cancer Chemother. Rep.* **1966**, *50*, 163–170.

23. Kruskal, W.H.; Wallis, W.A. Use of Ranks in One-Criterion Variance Analysis. *J. Am. Stat. Assoc.* **1952**, *47*, 583–621. [CrossRef]
24. Curatola, A.; Graglia, B.; Ferretti, S.; Covino, M.; Pansini, V.; Eftimiadi, G.; Chiaretti, A.; Gatto, A. The acute bronchiolitis rebound in children after COVID-19 restrictions: A retrospective, observational analysis. *Acta Bio Med. Atenei Parm.* **2023**, *94*, e2023031. [CrossRef]
25. Pruccoli, G.; Castagno, E.; Raffaldi, I.; Denina, M.; Barisone, E.; Baroero, L.; Timeus, F.; Rabbone, I.; Monzani, A.; Terragni, G.M.; et al. The Importance of RSV Epidemiological Surveillance: A Multicenter Observational Study of RSV Infection during the COVID-19 Pandemic. *Viruses* **2023**, *15*, 280. [CrossRef]
26. Olsen, S.J.; Winn, A.K.; Budd, A.P.; Prill, M.M.; Steel, J.; Midgley, C.M.; Kniss, K.; Burns, E.; Rowe, T.; Foust, A.; et al. Changes in Influenza and Other Respiratory Virus Activity During the COVID-19 Pandemic—United States, 2020–2021. *MMWR Morb. Mortal. Wkly. Rep.* **2021**, *70*, 1013–1019. [CrossRef]
27. Opek, M.W.; Yeshayahu, Y.; Glatman-Freedman, A.; Kaufman, Z.; Sorek, N.; Brosh-Nissimov, T. Delayed respiratory syncytial virus epidemic in children after relaxation of COVID-19 physical distancing measures, Ashdod, Israel, 2021. *Eurosurveillance* **2021**, *26*, 2100706. [CrossRef]
28. Barrezueta, L.B.; Zamorano, M.G.; López-Casillas, P.; Brezmes-Raposo, M.; Fernández, I.S.; Vázquez, M.d.l.A.P. Influencia de la pandemia COVID-19 sobre la epidemiología de la bronquiolitis aguda. *Enferm. Infecc. Microbiol. Clin.* **2023**, *41*, 348–351. [CrossRef]
29. Shanahan, K.H.; Monuteaux, M.C.; Bachur, R.G. Severity of Illness in Bronchiolitis Amid Unusual Seasonal Pattern During the COVID-19 Pandemic. *Hosp. Pediatr.* **2022**, *12*, e119–e123. [CrossRef] [PubMed]
30. Correction to: Outbreak of Respiratory Syncytial Virus Bronchiolitis in Italy. *Clin. Infect. Dis.* **2023**, *76*, 777–779. [CrossRef] [PubMed]
31. Manti, S.; Giallongo, A.; Parisi, G.F.; Papale, M.; Presti, S.; Bianco, M.L.; Spicuzza, L.; Leonardi, S. Impact of COVID-19 Pandemic and Lockdown on the Epidemiology of RSV-Mediated Bronchiolitis: Experience from Our Centre. *Children* **2022**, *9*, 1723. [CrossRef]
32. Presti, S.; Manti, S.; Gambilonghi, F.; Parisi, G.F.; Papale, M.; Leonardi, S. Comparative Analysis of Pediatric Hospitalizations during Two Consecutive Influenza and Respiratory Virus Seasons Post-Pandemic. *Viruses* **2023**, *15*, 1825. [CrossRef] [PubMed]
33. Nenna, R.; Matera, L.; Pierangeli, A.; Oliveto, G.; Viscido, A.; Petrarca, L.; La Regina, D.P.; Mancino, E.; Di Mattia, G.; Villani, A.; et al. First COVID-19 lockdown resulted in most respiratory viruses disappearing among hospitalised children, with the exception of rhinoviruses. *Acta Paediatr.* **2022**, *111*, 1399–1403. [CrossRef]
34. Ferrero, F.; Ossorio, M.F. Is there a place for bronchiolitis in the COVID-19 era? Lack of hospitalizations due to common respiratory viruses during the 2020 winter. *Pediatr. Pulmonol.* **2021**, *56*, 2372–2373. [CrossRef]
35. Wilder, J.L.; Parsons, C.R.; Growdon, A.S.; Toomey, S.L.; Mansbach, J.M. Pediatric Hospitalizations During the COVID-19 Pandemic. *Pediatrics* **2020**, *146*, e2020005983. [CrossRef]
36. Lumley, S.F.; Richens, N.; Lees, E.; Cregan, J.; Kalimeris, E.; Oakley, S.; Morgan, M.; Segal, S.; Dawson, M.; Walker, A.S.; et al. Changes in paediatric respiratory infections at a UK teaching hospital 2016–2021; impact of the SARS-CoV-2 pandemic. *J. Infect.* **2022**, *84*, 40–47. [CrossRef]
37. Koivisto, K.; Nieminen, T.; Mejias, A.; Gonzalez, C.C.; Ye, F.; Mertz, S.; Peeples, M.; Ramilo, O.; Saxén, H. Respiratory Syncytial Virus (RSV)–Specific Antibodies in Pregnant Women and Subsequent Risk of RSV Hospitalization in Young Infants. *J. Infect. Dis.* **2022**, *225*, 1189–1196. [CrossRef]
38. Manti, S.; Leonardi, S.; Rezaee, F.; Harford, T.J.; Perez, M.K.; Piedimonte, G. Effects of Vertical Transmission of Respiratory Viruses to the Offspring. *Front. Immunol.* **2022**, *13*, 853009. [CrossRef]
39. Hatter, L.; Eathorne, A.; Hills, T.; Bruce, P.; Beasley, R. Respiratory syncytial virus: Paying the immunity debt with interest. *Lancet Child Adolesc. Health* **2021**, *5*, e44–e45. [CrossRef]

viruses

Article

Clinical Differences between SARS-CoV-2 and RSV Infections in Infants: Findings from a Case–Control Study

Victor Daniel Miron [1,*], Raluca-Oana Raianu [1], Claudiu Filimon [1] and Mihai Craiu [1,2]

1 Carol Davila University of Medicine and Pharmacy, 050474 Bucharest, Romania
2 National Institute for Mother and Child Health "Alessandrescu-Rusescu", 020395 Bucharest, Romania
* Correspondence: mironvictordaniel@gmail.com

Abstract: Infants are a unique pediatric group due to their high hospitalization rates and unfavorable outcomes from acute infectious diseases. Understanding the clinical differences and aftereffects of SARS-CoV-2 in comparison to other prevalent viruses in this age group, like RSV, is crucial for effective management. We conducted a retrospective case–control study of infants hospitalized with SARS-CoV-2 or respiratory syncytial virus (RSV) infection in one year, in a tertiary pediatric hospital in Bucharest, Romania. A total of 188 infants were included in the analysis in a 1:1 ratio (94 with SARS-CoV-2 infection and 94 with RSV infection). Infants with COVID-19 were 10.2 times more likely to have fever ($p < 0.001$) and 2.4 times more likely to have diarrhea ($p = 0.016$). Conversely, infants with RSV were 2.5 times more likely to have a cough ($p < 0.001$), 3.0 times more likely to have nasal congestion ($p < 0.001$), and 14.7 times more likely to present with dyspnea ($p < 0.001$). Increased lymphocyte count was more common in infants with RSV ($p = 0.008$), while lymphopenia was more frequent in infants with SARS-CoV-2 ($p = 0.011$). The median length of hospital stay was one day longer in infants with RSV infection (5 days vs. 4 days). Overall, infants with RSV infection had a 27.3-fold increased risk of developing respiratory failure ($p < 0.001$), while infants with COVID-19 had a 5.8-fold increased risk of laryngitis ($p = 0.003$). Our findings suggest that infants with SARS-CoV-2 infection may present with polymorphic symptoms, mostly dominated by fever, whereas infants with RSV often present with respiratory symptoms. Laboratory differentiation between the two infections is challenging; therefore, the use of rapid antigen or molecular diagnostic tests is crucial for accurate diagnosis, epidemiologically appropriate measures, and effective management. Continued surveillance of both viruses in infants, and beyond, and the implementation of specific control measures are needed to mitigate their impact on this vulnerable pediatric group.

Keywords: infant; SARS-CoV-2; COVID-19; RSV; clinical features; respiratory failure; laryngitis

Citation: Miron, V.D.; Raianu, R.-O.; Filimon, C.; Craiu, M. Clinical Differences between SARS-CoV-2 and RSV Infections in Infants: Findings from a Case–Control Study. *Viruses* **2024**, *16*, 63. https://doi.org/10.3390/v16010063

Academic Editors: Emanuele Castagno and Irene Raffaldi

Received: 12 December 2023
Revised: 27 December 2023
Accepted: 29 December 2023
Published: 30 December 2023

1. Introduction

The first year of life is an extremely important period in a child's development. Any external factor, especially acute illnesses, can interfere with and disrupt the developmental process. In addition, an untrained and still evolving immune system puts infants at risk of an unfavorable outcome in the event of an acute infectious disease [1]. Therefore, this group of children requires close monitoring to prevent potential complications.

Based on experience with influenza and respiratory syncytial virus (RSV) infections, where young children are at risk of hospitalization and potentially severe outcomes, the emergence of circulating SARS-CoV-2 was viewed with special concern for this group of children [2].

More than three years after the onset of the COVID-19 pandemic, it has been observed that SARS-CoV-2 infection has had a significant impact on the pediatric population, and although the majority of cases have had favorable outcomes [3], there is still a need for further research; there have also been intensive care unit (ICU) admissions and deaths associated with COVID-19. In an analysis of causes of death among children and young people aged 0 to 19 years in the United States, SARS-CoV-2 infection was included in the top ten, precisely ranking eighth [4].

The true impact of SARS-CoV-2 infection in infants has not been fully quantified and characterized, as data for this age group are clearly lacking. A good way to fully characterize COVID-19 in infants is to compare it with other viral infections with a significant impact on this population, in particular RSV infection. RSV is known to be the leading cause of hospitalization and bronchiolitis in infants [5].

In this context, we aimed to conduct a case–control study describing the clinical course and evolution of COVID-19 in hospitalized infants by conducting a comparative analysis with infants with RSV infection.

2. Methods

We conducted a retrospective case–control study among infants hospitalized with SARS-CoV-2 or RSV infection at the National Institute for Mother and Child Health "Alessandrescu-Rusescu" (NIMCH) over a 12-month period from April 2021 to March 2022 (Delta and Omicron SARS-CoV-2 variants).

NIMCH is one of four university children's hospitals in Bucharest, the main city of Romania. Approximately 10,000 children are admitted to the hospital each year, and more than 60,000 children are seen in the outpatient and/or emergency departments. At the beginning of the COVID-19 pandemic, NIMCH was designated as the hospital to treat non-COVID-19 cases, mainly respiratory and digestive pathology. Since April 2021, an eight-bed ward has been opened in the hospital for the admission of children with SARS-CoV-2 infection.

Cases in this study were specifically identified as infants who were hospitalized during the study period and confirmed to have SARS-CoV-2 infection by reverse transcription-polymerase chain reaction (RT-PCR) testing. We included all cases consecutively, excluding children older than 1 year at the time of hospitalization, those with incomplete electronic data, and those with other co-infections as evidenced by rapid tests, multiplex RT-PCR, or cultures. This approach was taken to ensure that our study focused exclusively on cases infected with SARS-CoV-2 only.

Controls were defined as infants with RSV infection confirmed by rapid antigen or multiplex RT-PCR testing who were hospitalized during the study period. We matched the controls to the cases consecutively by sex and age group. Exclusions were similar to the cases: children over 1 years old at hospitalization, those lacking complete electronic data, and those with other proven co-infections (similar to that described above for the cases).

The case:control ratio was set at 1:1. The equal division of the study participants into these two groups allowed for a balanced comparison, ensuring that each group was adequately represented in the analysis. This parity is essential for a fair and accurate comparison between the two infections, providing insights into their similarities and differences in terms of symptoms, severity, and overall impact on the health of infants.

For each infant identified as a case or control in the study, a complete set of data including demographic and clinical data, laboratory parameter values, and outcomes was collected. In order to facilitate a more detailed and nuanced analysis, the infants in our study were categorized into five distinct age groups. This segmentation was based on their age at the time of their involvement in the study: newborn (0–28 days), 1–3 months, 4–6 months, 7–9 months, and 10–12 months.

A statistical analysis was performed using IBM SPSS Statistics software, version 25 (IBM Corp., Armonk, NY, USA). The level of statistical significance was set at $p < 0.05$. Categorical data were compared using the chi-squar (χ^2) test, and risk was reported as the odds ratio (OR) along with the 95% confidence interval (95%CI). Continuous variables were analyzed using the Mann–Whitney-U (U) test with z-value (z) and effect size (r) because their distribution was non-Gaussian.

3. Results

A total of 188 infants were included in the analysis, equally divided into two groups of 94 each, one with SARS-CoV-2 infection and the other with RSV infection. The demographic characteristics of both groups were comparable and are summarized in Table 1.

Table 1. Demographic characteristics of the groups analyzed.

Characteristic	SARS-CoV-2 Infection, *n* (%)	RSV Infection, *n* (%)	Total, *n* (%)
Sex			
Male	54 (57.4)	54 (57.4)	108 (57.4)
Female	40 (42.6)	40 (42.6)	80 (42.6)
Age group			
0–28 days	9 (9.6)	9 (9.6)	18 (9.6)
1–3 months	45 (47.9)	45 (47.9)	90 (47.9)
4–6 months	19 (20.2)	19 (20.2)	38 (20.2)
7–9 months	13 (13.8)	13 (13.8)	26 (13.8)
10–12 months	8 (8.5)	8 (8.5)	16 (8.5)
Median age (IQR), in months	3 (IQR: 2.0, 5.8)	3 (IQR: 1.0, 6.0)	3 (IQR: 1.0, 6.0)

RSV—respiratory syncytial virus.

An epidemiological context was more commonly identified among infants with COVID-19 (51.1%, $n = 48$) than those with RSV infection (36.2%, $n = 34$, $p = 0.039$, $\chi^2 = 61.3$, OR = 0.5, 95%CI: 0.3–0.97), particularly in the 1–3 and 4–6 months age groups (see Supplementary Materials).

A total of 22 (11.7%) infants included in the study had at least one chronic disease. Of these infants, 9 had SARS-CoV-2 infection and 13 had RSV infection ($p = 0.364$). Congenital heart diseases were most commonly identified (10 infants, 4 COVID-19, and 6 RSV), followed by genetic diseases (8 infants, 5 COVID-19, and 3 RSV), neurological diseases (8 infants, 3 COVID-19, and 5 RSV), and pulmonary diseases (4 infants, all with RSV infection).

Infants with SARS-CoV-2 infection generally presented to the hospital earlier than those with RSV infection. The median hospital presentation was 1 day for COVID-19 (IQR: 1.0–2.3 days) versus 3 days for RSV infection (IQR: 2.0–4.0 days), $p < 0.001$, $z = -6.997$, $U = 1878.5$.

Significant clinical differences were observed between the two infections. Infants with COVID-19 were 10.2 times more likely to have fever ($p < 0.001$, OR = 10.2) and 2.4 times more likely to have diarrhea ($p = 0.016$, OR = 2.4). Conversely, infants with RSV were 2.5 times more likely to have a cough ($p < 0.001$, OR = 0.5), 3.0 times more likely to have nasal congestion ($p < 0.001$, OR = 0.3), and 14.7 times more likely to present with dyspnea ($p < 0.001$, OR = 0.07) (Table 2). In the age group analysis (Supplementary Materials), we found that cough and nasal congestion were significant in infants younger than 6 months with RSV, whereas fever was significantly predominant in infants aged 1–9 months with COVID-19.

In terms of laboratory changes, an increase in monocytes was the most common finding in both groups (83.0% and 80.9%, respectively, $p = 0.705$, Table 3). However, increased lymphocyte counts were more common in infants with RSV infection ($p = 0.008$, OR = 0.5), while decreased lymphocyte counts were more frequent in infants with SARS-CoV-2 infection ($p = 0.011$, OR = 3.1). Additionally, infants with COVID-19 were 2.5 times more likely to have an inflammatory syndrome characterized by elevated C-reactive protein levels compared to those with RSV infection ($p = 0.007$, OR = 2.6). Table 3 summarizes the laboratory changes analyzed between the two groups. By age group, the trend of changes was similar to those highlighted above, and they are presented in the Supplementary Material.

Table 2. Comparison of the clinical characteristics between the groups.

Clinical Characteristic	SARS-CoV-2 Infection, N = 94, *n* (%)	RSV Infection, N = 94, *n* (%)	Statistical Analysis
Fever	77 (81.9)	29 (30.9)	$p < 0.001$, $\chi^2 = 49.8$, OR = 10.2, 95%CI: 5.12–20.11
Maximum temperature, °C, median (IQR)	38.8 (IQR: 38.5–39.3)	38.5 (IQR: 38.3–38.9)	$p = 0.039$, U = 621, z = −2.061, r = 0.150
Cough	64 (68.1)	94 (100.0)	$p < 0.001$, $\chi^2 = 35.7$, OR = 0.5, 95%CI: 0.25–0.82
Nasal congestion	45 (47.9)	69 (73.4)	$p < 0.001$, $\chi^2 = 12.8$, OR = 0.3, 95%CI: 0.2–0.6
Dyspnea	13 (13.8)	66 (70.2)	$p < 0.001$, $\chi^2 = 61.3$, OR = 0.07, 95%CI: 0.03–0.14
Diarrhea	29 (30.9)	15 (16.0)	$p = 0.016$, $\chi^2 = 5.8$, OR = 2.4, 95%CI: 1.16–4.75
Vomiting	12 (12.8)	14 (14.9)	$p = 0.673$, $\chi^2 = 0.2$, OR = 0.8, 95%CI: 0.37–1.91
Malaise	15 (16.0)	31 (33.0)	$p = 0.007$, $\chi^2 = 7.3$, OR = 0.4, 95%CI: 0.19–0.78

RSV—respiratory syncytial virus.

Table 3. Comparison of the laboratory parameters between the groups.

Laboratory Parameter	SARS-CoV-2 Infection, N = 94, *n* (%)	RSV Infection, N = 94, *n* (%)	Statistical Analysis
Increased white blood cell count	18 (19.1)	18 (19.1)	$p = 1.000$, $\chi^2 = 0.0$, OR = 1.0, 95%CI: 0.48–2.07
Decreased white blood cell count	10 (10.6)	7 (7.4)	$p = 0.446$, $\chi^2 = 0.6$, OR = 1.5, 95%CI: 0.53–4.07
Increased neutrophil count	5 (5.3)	1 (1.1)	$p = 0.097$, $\chi^2 = 2.8$, OR = 5.2, 95%CI: 0.60–45.61
Decreased neutrophil count	24 (25.5)	18 (19.1)	$p = 0.293$, $\chi^2 = 1.1$, OR = 1.5, 95%CI: 0.73–2.89
Increased monocyte count	78 (83.0)	76 (80.9)	$p = 0.705$, $\chi^2 = 0.1$, OR = 1.2, 95%CI: 0.55–2.43
Increased lymphocyte count	33 (35.1)	51 (54.3)	$p = 0.008$, $\chi^2 = 6.9$, OR = 0.5, 95%CI: 0.25–0.82
Decreased lymphocyte count	19 (20.2)	7 (7.4)	$p = 0.011$, $\chi^2 = 6.4$, OR = 3.1, 95%CI: 1.26–7.90
Increased C-reactive protein	31 (33.0)	15 (16.0)	$p = 0.007$, $\chi^2 = 7.4$, OR = 2.6, 95%CI: 1.29–5.22
Increased AST	12 (12.8)	48 (51.1)	$p < 0.001$, $\chi^2 = 31.7$, OR = 0.14, 95%CI: 0.07–0.29
Increased ALT	8 (8.5)	9 (9.6)	$p = 0.799$ $\chi^2 = 0.1$, OR = 0.8, 95%CI: 0.32–2.38
Increased urea	7 (7.4)	0 (0.0)	NA
Increased creatinine	1 (1.1)	0 (0.0)	NA
Chest X-ray			
Normal	11 (57.9)	2 (16.7)	$p = 0.023$, $\chi^2 = 5.1$, OR = 0.3, 95%CI: 0.05–0.8
Interstitial pneumonia	8 (47.1)	9 (52.9)	$p = 0.806$, $\chi^2 = 0.1$, OR = 0.9, 95%CI: 0.32–2.38
Lung consolidation	0 (0.0)	1 (8.3)	NA

AST—aspartate aminotransferase; ALT—alanine aminotransferase; NA—not applicable; RSV—respiratory syncytial virus.

A reduced number of only 31 infants (31/188, 16.5%) required imaging investigations such as chest X-ray (19 infants (20.2%) with COVID-19 and 12 infants (12.8%) with RSV infection, $p = 0.169$). In 54.8% of cases (17/31), these infants showed interstitial pneumonia, and in 41.9% (13/31), they were normal without any pathological changes. Normal appearance was more common in SARS-CoV-2 infection (57.9% vs. 16.7%, $p < 0.001$, OR = 6.9), Table 3.

The median length of hospital stay was 1 day longer in infants with RSV infection: 5 days (IQR:3, 7 days) versus 4 days (IQR:3, 7 days) for COVID-19, $p = 0.079$, $z = -1.756$, U = 3768.0. The presence of acute respiratory failure in infants with RSV significantly increased the median length of hospital stay by 3 days ($p < 0.001$). The outcome was favorable for all cases, with no deaths in either group. Overall, infants with RSV infection had a 27.3-fold increased risk of respiratory failure (2.1%, $n = 2$ vs. 37.2%, $n = 35$, $p < 0.001$, $\chi^2 = 36.6$, OR = 0.04, 95%CI: 0.008–0.16), whereas infants with COVID-19 had a 5.8-fold increased risk of acute laryngitis (16.0%, $n = 15$ vs. 3.2%, $n = 3$, $p = 0.003$, $\chi^2 = 8.8$, OR = 5.8, 95%CI: 1.6–20.6) (Figure 1). The incidence of acute dehydration syndrome was relatively similar between the two groups (56.4%, $n = 53$ for SARS-CoV-2 vs. 47.9%, $n = 45$ for RSV, $p = 0.243$). Only five cases were complicated by acute otitis media, one case in an infant with COVID-19 and four cases in children with RSV ($p = 0.174$).

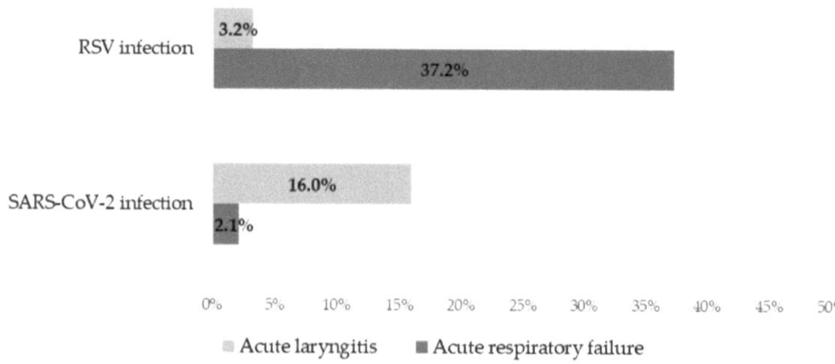

Figure 1. Rates of acute respiratory failure and acute laryngitis by group.

4. Discussion

We conducted a case–control analysis to highlight the characteristics of SARS-CoV-2 infection in contrast to RSV infection in hospitalized infants over a period of one year in a pediatric teaching hospital in Bucharest. Respiratory infections are responsible for an increased morbidity and mortality rate among the pediatric population. RSV is one of the most incriminated viruses involved, among young children, especially infants [6]. It has been observed that 45% of hospitalizations due to RSV are reported in children under 6 months of age, which is associated with a risk of unfavorable outcomes with respiratory failure [7]. Studies have also shown that the majority of infants with RSV requiring hospitalization were previously healthy and born at term [8], while premature children or children with certain underlying medical conditions are at higher risk of ICU admission [9]. In contrast, SARS-CoV-2 infection in children resulted in mild-to-moderate disease. Hospitalizations have been reported mainly in young children and infants, mainly for medical follow-up, as the natural course of the disease is benign [10]. However, ICU admissions were also reported among infants with COVID-19, but the admission rate was less than 2% [11].

A European (including data from our country) surveillance study of epidemiology aspects during the SARS-CoV-2 pandemic [12] documented a decrease in RSV circulation to a low rate of positivity, 1% of specimens, from more than 21,000 primary care sentinel surveillance centers. Only France and Switzerland documented clear waves of RSV activity, and France, Germany, and Slovenia recorded an early start to their seasonal 2021–2022 circulation compared with the average starting week in pre-COVID-19 pandemic seasons [12]. This important change in circulation paradigm was also documented in other parts of the world, where RSV circulates during the winter months, such as in Australia and New Zealand [13] and South Africa [14].

One particular aspect of RSV circulation in our country, during the study period, was the lack of prophylactic measures for severe RSV infections in vulnerable populations. Former premature children are at risk of developing respiratory failure during an RSV episode, as described by pre-pandemic papers from our country [15]. From 2022, palivizumab (monoclonal anti-RSV antibodies) became available again, after more than a decade of unattainability, for these children (premature infants with a shorter gestational age than 35 weeks, children younger than 2 years of age with bronchopulmonary dysplasia or with severe congenital heart disease, and infants with cystic fibrosis, neuromuscular disorders or immune-deficit syndromes [16]). Protection for severe RSV infection, in Romanian children, will be increased in the near future through the use of nirsevimab, after European Medicines Agency (EMA) approval in September 2022 [17].

In our analysis, we included cases of the COVID-19 variants Delta and Omicron, when both in Romania [10], and other countries [18], the number of cases of SARS-CoV-2 infection in children was increasing. Infants with RSV infection were selected from the same period by matching cases according to sex and age for robust conclusions regarding our data.

The specific prevalence of the SARS-CoV-2 virus within our country can likely be attributed to multiple contributing factors. One significant aspect is the relatively low rate of vaccination among eligible individuals, with only 42.2% of the general population and 50.7% of those aged 18 and above being vaccinated [19]. These figures are notably lower compared to the European Union/European Economic Area (EU/EEA) average, which stands at 73%. This underwhelming vaccination uptake is indicative of deeper societal issues, such as a widespread lack of trust in governmental institutions, the influence of misinformation campaigns, inadequate infrastructure in rural areas, and a deficiency in effective vaccine education and outreach efforts [20–22]. These factors have collectively played a role in shaping the country's response to the pandemic and the virus's circulation.

Infants with COVID-19 most commonly presented to hospital with fever (81.9%), with or without respiratory symptoms such as cough (68.1%) or nasal congestion (47.9%). One third of cases had gastrointestinal symptoms, with diarrhea being the most common. In another Romanian study of 613 infants hospitalized with SARS-CoV-2 infection, fever was also the main symptom (96.4%), followed by cough (64.8%), and diarrhea was present in 37.5% of children [10]. In an analysis from Poland of 300 hospitalized cases of COVID-19 in infants, similar data were reported with a frequency of 77% fever, 40% cough, and 24% diarrhea [23].

In contrast, the infants with RSV in the study had a clinical picture dominated by respiratory symptoms, with cough present in all cases, followed by nasal congestion (73.4%). A significant proportion of infants (70.2%) had dyspnea, while fever was present in one third of patients with RSV. In a retrospective study of children up to 5 years of age with RSV in Germany, the majority of hospitalized patients were infants and respiratory manifestations dominated the clinical picture, with dyspnea present in 79.6% of infants aged 0–6 months and 81.3% of infants aged 6 months to 1 year [24].

Our analysis shows clear differences in the clinical presentation of the two infections in infants. While RSV infection is more likely to present as a clinical respiratory syndrome with cough and/or dyspnea, SARS-CoV-2 infection may present as a polymorphic picture dominated by fever in most cases. Similar data were reported in a review by Fedorczak et al. for infants younger than 3 months only [25].

In our study, we observed that infants with SARS-CoV-2 infection presented to hospital earlier than those with RSV infection. This could be explained in particular by the more frequent presence of fever in those with COVID-19, as fever is known to be one of the clinical symptoms most feared by parents, especially in this age group. Hatmann et al. had similar results to ours, showing hospital presentation for children with RSV on average 3 days after the onset of symptoms [24].

In terms of laboratory investigations, the most common change described in both groups was an increase in monocytes. In particular, lymphocytosis was more common in infants with RSV and lymphopenia was more common in infants with SARS-CoV-2. Brüssow et al. [26] showed that the decreased lymphocyte count associated with SARS-CoV-2 infection could be explained by the inhibitory effect of activated cytokines. This also overlaps with our finding that elevated C-reactive protein values were more common in infants with COVID-19.

Our findings indicated a distinct difference in radiological presentations between infants suffering from COVID-19 and those with RSV. Specifically, infants afflicted with COVID-19 tended to exhibit normal imaging results more frequently. In contrast, those diagnosed with RSV were more prone to show signs of interstitial pneumonia in their radiological evaluations. This divergence in radiological findings could potentially be attributed to the greater likelihood of lower airway involvement in infants with RSV, as compared to those with COVID-19. This aspect of our study highlights the variances in how these two viral infections manifest in pediatric patients, particularly in terms of their impact on the respiratory system as observed through imaging studies.

COVID-19 infants had a shorter hospital stay than those with RSV, in whom we observed that the presence of acute respiratory failure increased the length of hospitalization by at least 3 days. Our results are consistent with data from other studies that have highlighted the need for longer hospital stays and more complex management among children with RSV compared to SARS-CoV-2 [25,27]. In our analysis focusing on the length of hospital stay in different age groups, we found that the median length of hospital stay did not vary significantly between groups. This observation suggests a relatively consistent pattern in the length of hospital stay across different age groups of infants. However, it is noteworthy that the shortest median length of hospital stay was specifically observed in the group of newborns infected with COVID-19. This aspect of our analysis provides valuable insights into how the duration of hospitalization for COVID-19 may differ in the earliest stages of life compared to older infants.

The outcome of all infants was positive and there were no deaths. Infants infected with RSV had a higher risk of developing respiratory failure, while those infected with COVID-19 had a higher risk of laryngitis. The results from other trials also suggest a favorable outcome for both infections [25,27], but the most common complication observed in the short term was the suspicion of bacterial superinfection [25].

In the current context of COVID-19 pandemic development, vigilant surveillance of RSV and SARS-CoV-2, along with influenza and other respiratory viruses, is extremely important [28,29]. These viruses pose significant public health challenges in terms of high numbers of patients seeking medical care, high rates of hospitalization and high numbers of deaths. In addition to its known impact on infants and young children, RSV is now known to be an important respiratory pathogen in older adults, particularly those with associated chronic diseases, and can lead to decompensation and progression with severe complications. At the same time, SARS-CoV-2 has demonstrated its capacity for rapid mutation, leading to various circulant variants with differing levels of transmissibility and severity. Persistent surveillance is crucial for the early detection of new variants, which is essential for timely public health responses, including updates to vaccines and treatment protocols.

Surveillance also plays a critical role in understanding the epidemiology of these viruses, such as transmission dynamics, infection rates, and population immunity. This information is vital for guiding public health policies, including vaccination strategies, social distancing measures, and resource allocation for healthcare facilities. Concurrent surveillance of RSV and SARS-CoV-2 is vital due to the potential for co-infections, which can exacerbate the severity of illness and complicate treatment strategies. Such surveillance helps in preparing healthcare systems for potential surges in cases, ensuring adequate medical supplies, and informing the public about preventive measures.

Although our study provides valuable insights, it has several limitations, mainly due to its retrospective design and the lack of extended follow-up of the infants included in our analysis. Another notable limitation of our study is the relatively small sample size of patients. A limited number of participants may reduce the statistical power of the study, making it more difficult to detect significant differences or draw robust conclusions. This small sample size may also limit the generalizability of our findings to the wider population. However, it is important to highlight a strength of our study, which is the careful matching of SARS-CoV-2 and RSV cases. By rigorously matching these cases, our study provides a clearer and more detailed comparison of the differences between the two infections in infants. This aspect of our methodology increases the reliability of our findings in understanding how these infections differ, particularly in their clinical presentation, severity, and short-term outcomes in this vulnerable age group.

5. Conclusions

The findings of this study suggest that infants with SARS-CoV-2 could be discriminated by clinical phenotyping from those with RSV infection, in the initial triage approach in an emergency department in order to have optimal hospital management, through the use of a tailored viral detection strategy, especially in a low-resource setting where monoclonal antibodies or maternal RSV vaccine prophylaxis are not available. Infants with SARS-CoV-2 infection may have polymorphic symptoms, mostly dominated by fever, whereas infants with RSV often have a cough and/or shortness of breath. Continued surveillance of both viruses in infants, and beyond, and the implementation of specific control measures are needed to mitigate their impact on this vulnerable pediatric group, according to dynamic, changing post-COVID-19 pandemic seasonal circulation.

Supplementary Materials: The following are available online at: https://www.mdpi.com/article/10.3390/v16010063/s1. Table S1. Patient characteristics by age group and type of infection.

Author Contributions: Conceptualization, V.D.M. and R.-O.R.; methodology, V.D.M.; software, V.D.M., C.F. and R.-O.R.; validation, V.D.M., C.F. and R.-O.R.; formal analysis, V.D.M. and R.-O.R.; investigation, V.D.M. and R.-O.R.; data curation, V.D.M., C.F. and R.-O.R.; writing—original draft preparation, V.D.M. and C.F.; writing—review and editing, V.D.M., C.F., R.-O.R. and M.C.; supervision, M.C. All authors have read and agreed to the published version of the manuscript.

Funding: This research received no external funding.

Institutional Review Board Statement: The study was conducted in accordance with the Declaration of Helsinki and approved by Ethics Committee of National Institute for Mother and Child Health Alessandrescu-Rusescu, Bucharest.

Informed Consent Statement: Not applicable.

Data Availability Statement: The datasets generated and analysed during the current study are available from the corresponding author upon reasonable request.

Conflicts of Interest: The authors declare no conflicts of interest.

References

1. Simon, A.K.; Hollander, G.A.; McMichael, A. Evolution of the immune system in humans from infancy to old age. *Proc. R. Soc. B Biol. Sci.* **2015**, *282*, 20143085. [CrossRef] [PubMed]
2. Miron, V.D.; Gunșahin, D.; Filimon, C.; Bar, G.; Craiu, M. Pediatric Emergencies and Hospital Admissions in the First Six Months of the COVID-19 Pandemic in a Tertiary Children's Hospital in Romania. *Children* **2022**, *9*, 513. [CrossRef] [PubMed]
3. Vladescu, C.; Ciutan, M.; Rafila, A. In-hospital admissions and deaths in the context of the COVID-19 pandemic, in Romania. *Germs* **2022**, *12*, 169–179. [CrossRef] [PubMed]
4. Flaxman, S.; Whittaker, C.; Semenova, E.; Rashid, T.; Parks, R.M.; Blenkinsop, A.; Unwin, H.J.T.; Mishra, S.; Bhatt, S.; Gurdasani, D.; et al. Assessment of COVID-19 as the Underlying Cause of Death Among Children and Young People Aged 0 to 19 Years in the US. *JAMA Netw. Open* **2023**, *6*, e2253590. [CrossRef]

5. Rodriguez-Fernandez, R.; González-Sánchez, M.I.; Perez-Moreno, J.; González-Martínez, F.; Navazo, S.d.l.M.; Mejias, A.; Ramilo, O. Age and respiratory syncytial virus etiology in bronchiolitis clinical outcomes. *J. Allergy Clin. Immunol. Glob.* **2022**, *1*, 91–98. [CrossRef]
6. Suryadevara, M.; Domachowske, J.B. Epidemiology and Seasonality of Childhood Respiratory Syncytial Virus Infections in the Tropics. *Viruses* **2021**, *13*, 696. [CrossRef]
7. Shi, T.; McAllister, D.A.; O'Brien, K.L.; Simoes, E.A.F.; Madhi, S.A.; Gessner, B.D.; Polack, F.P.; Balsells, E.; Acacio, S.; Aguayo, C.; et al. Global, regional, and national disease burden estimates of acute lower respiratory infections due to respiratory syncytial virus in young children in 2015: A systematic review and modelling study. *Lancet* **2017**, *390*, 946–958. [CrossRef]
8. Rha, B.; Curns, A.T.; Lively, J.Y.; Campbell, A.P.; Englund, J.A.; Boom, J.A.; Azimi, P.H.; Weinberg, G.A.; Staat, M.A.; Selvarangan, R.; et al. Respiratory Syncytial Virus–Associated Hospitalizations Among Young Children: 2015–2016. *Pediatrics* **2020**, *146*, e20193611. [CrossRef]
9. Pham, H.; Thompson, J.; Wurzel, D.; Duke, T. Ten years of severe respiratory syncytial virus infections in a tertiary paediatric intensive care unit. *J. Paediatr. Child Health* **2020**, *56*, 61–67. [CrossRef]
10. Draganescu, A.C.; Miron, V.D.; Sandulescu, O.; Bilasco, A.; Streinu-Cercel, A.; Sandu, R.G.; Marinescu, A.; Gunșahin, D.; Hoffmann, K.L.; Horobeț, D.S.; et al. Omicron in Infants-Respiratory or Digestive Disease? *Diagnostics* **2023**, *13*, 421. [CrossRef]
11. Al Dhaheri, F.A.; El Dannan, H.; Hashim, M.J.; Alshehi, S.; Al-Jburi, F.; Antali, A.; Al Jasmi, N.; Al Khouri, S.; Al Hajjar, M.; Abbas, T.; et al. Clinical Outcomes of COVID-19 in Newborns and Infants: A Multicenter Experience of 576 Cases. *Pediatr. Infect. Dis. J.* **2023**, *42*, 515–519. [CrossRef] [PubMed]
12. Meslé, M.M.I.; Sinnathamby, M.; Mook, P.; Group WHOERRN; Pebody, R.; Lakhani, A. The WHO European Region Respiratory Network Group Seasonal and inter-seasonal RSV activity in the European Region during the COVID-19 pandemic from autumn 2020 to summer 2022. *Influ. Other Respir. Viruses* **2023**, *17*, e13219. [CrossRef] [PubMed]
13. Binns, E.; Koenraads, M.; Hristeva, L.; Flamant, A.; Baier-Grabner, S.; Loi, M.; Lempainen, J.; Osterheld, E.; Ramly, B.; Chakakala-Chaziya, J.; et al. Influenza and respiratory syncytial virus during the COVID-19 pandemic: Time for a new paradigm? *Pediatr. Pulmonol.* **2022**, *57*, 38–42. [CrossRef] [PubMed]
14. Odumade, O.A.; van Haren, S.D.; Angelidou, A. Implications of the Severe Acute Respiratory Syndrome Coronavirus 2 (SARS-CoV-2) Pandemic on the Epidemiology of Pediatric Respiratory Syncytial Virus Infection. *Clin. Infect. Dis.* **2022**, *75* (Suppl. S1), S130–S135. [CrossRef] [PubMed]
15. Bogdan, R.D.; Rusu, L.; Toma, A.I.; Nastase, L. Respiratory Outcome of the Former Premature Infants. *J. Med. Life* **2019**, *12*, 381–394. [CrossRef] [PubMed]
16. Romanian Ministry of Health. National RSV Prevention Protocol in Romania (Protocol Terapeutic Corespunzator Pozitiei nr. 284, cod(J06BB16): Dci Palivizumabum): Monitorul Oficial al Romaniei; 2022. Partea I, Nr. 1070 bis/4.XI.2022. Available online: https://monitoruloficial.ro/Monitorul-Oficial--PI--1070--2022.html (accessed on 12 December 2023).
17. European Medicines Agency. Beyfortus. 2022. Available online: https://www.ema.europa.eu/en/medicines/human/EPAR/beyfortus (accessed on 15 December 2023).
18. Nathanielsz, J.; Toh, Z.Q.; Do, L.A.H.; Mulholland, K.; Licciardi, P.V. SARS-CoV-2 infection in children and implications for vaccination. *Pediatr. Res.* **2023**, *93*, 1177–1187. [CrossRef] [PubMed]
19. European Centre for Disease Prevention and Control. Cumulative Uptake (%) of the Primary Course in the Total Population in Romania as of 5 October 2023. 2023. Available online: https://vaccinetracker.ecdc.europa.eu/public/extensions/COVID-19/vaccine-tracker.html#uptake-tab (accessed on 24 December 2023).
20. Dascalu, S. The Successes and Failures of the Initial COVID-19 Pandemic Response in Romania. *Front. Public Health* **2020**, *8*, 344. [CrossRef]
21. Ilea, C.D.N.; Daina, M.D.; Venter, A.C.; Șuteu, C.L.; Sabău, M.; Badau, D.; Daina, L.G. The Motivation of Medical Staff and the Work Inter-estedness in the Context of the COVID-19 Pandemic, in a Tertiary Hospital in Romania. *Healthcare* **2023**, *11*, 813. [CrossRef]
22. Gherheș, V.; Cernicova-Buca, M.; Fărcașiu, M.A. Public Engagement with Romanian Government Social Media Accounts during the COVID-19 Pandemic. *Int. J. Environ. Res. Public Health* **2023**, *20*, 2372. [CrossRef]
23. Sobolewska-Pilarczyk, M.; Pokorska-Śpiewak, M.; Stachowiak, A.; Marczyńska, M.; Talarek, E.; Ołdakowska, A.; Kucharek, I.; Sybilski, A.; Mania, A.; Figlerowicz, M.; et al. COVID-19 infections in infants. *Sci. Rep.* **2022**, *12*, 7765. [CrossRef]
24. Hartmann, K.; Liese, J.G.; Kemmling, D.; Prifert, C.; Weissbrich, B.; Thilakarathne, P.; Diels, J.; Weber, K.; Streng, A. Clinical burden of respiratory syncytial virus in hospitalized children aged$\leq$ 5 years (INSPIRE Study). *J. Infect. Dis.* **2022**, *226*, 386–395. [CrossRef] [PubMed]
25. Fedorczak, A.; Zielinska, N.; Nosek-Wasilewska, P.; Mikolajczyk, K.; Lisiak, J.; Zeman, K.; Tkaczyk, M. Comparison of COVID-19 and RSV Infection Courses in Infants and Children under 36 Months Hospitalized in Paediatric Department in Fall and Winter Season 2021/2022. *J. Clin. Med.* **2022**, *11*, 7088. [CrossRef] [PubMed]
26. Brussow, H. Immunology of COVID-19. *Environ. Microbiol.* **2020**, *22*, 4895–4908. [CrossRef] [PubMed]
27. Meyer, M.; Ruebsteck, E.; Eifinger, F.; Klein, F.; Oberthuer, A.; van Koningsbruggen-Rietschel, S.; Huenseler, C.; Weber, L.T. Morbidity of Respiratory Syncytial Virus (RSV) Infections: RSV Compared with Severe Acute Respiratory Syndrome Coronavirus 2 Infections in Chil-dren Aged 0–4 Years in Cologne, Germany. *J. Infect. Dis.* **2022**, *226*, 2050. [CrossRef]

28. Tarzjani, S.D.; Kamalzadeh, S.; Moghadam, M.T.; Ashoobi, M.T. Clinical challenge of co-infection of SARS-CoV-2 with influenza during the influenza circulation season: Suggestions for prevention. *Germs* **2023**, *13*, 188–191. [CrossRef]
29. Strouthou, E.; Karageorgos, S.A.; Christaki, E.; Agouridis, A.P.; Tsioutis, C. Medical students' attitudes and perceptions of influenza and SARS-CoV-2 vaccination in Cyprus. *Germs* **2022**, *12*, 180–194. [CrossRef]

viruses

MDPI

Brief Report

Characteristics of Respiratory Syncytial Virus Infections in Children in the Post-COVID Seasons: A Northern Italy Hospital Experience

Davide Treggiari [1,*], Chiara Pomari [2], Giorgio Zavarise [3], Chiara Piubelli [1,*], Fabio Formenti [1] and Francesca Perandin [1]

[1] Department of Tropical, Infectious Diseases and Microbiology, IRCCS Sacro Cuore-Don Calabria Hospital, 37024 Negrar di Valpolicella, Verona, Italy; fabio.formenti@sacrocuore.it (F.F.); francesca.perandin@sacrocuore.it (F.P.)
[2] Andrus Center, University of Southern California, Los Angeles, CA 90089, USA; chiara.pomari@icloud.com
[3] Department of Pediatrics, IRCCS Sacro Cuore-Don Calabria Hospital, 37024 Negrar di Valpolicella, Verona, Italy; giorgio.zavarise@sacrocuore.it
* Correspondence: davide.treggiari@sacrocuore.it (D.T.); chiara.piubelli@sacrocuore.it (C.P.); Tel.: +39-045-601-3706 (C.P.)

Abstract: Background: Public health measures for COVID-19 mitigation influenced the circulation of Respiratory Syncytial Virus (RSV) during the 2020–2021 winter season. In the following autumn, an unprecedented resurgence of RSV occurred. Our study monitored RSV pediatric infections one and two years after the relaxation of containment measures for the COVID-19 pandemic. Methods: We analyzed diagnostic molecular data for SARS-CoV-2, flu, and RSV infections and clinical data from children with respiratory symptoms referring to our hospital during the 2021–2022 and 2022–2023 seasons. Results: In the 2021–2022 season, the number of RSV-affected children was very high, especially for babies <1 year. The outbreak appeared in a shorter interval of time, with a high clinical severity. In the 2022–23 season, a reduced number of infected pediatric patients were detected, with a similar hospitalization rate (46% vs. 40%), and RSV accounted for 12% of the infections. Coinfections were observed in age <2 years. In RSV patients, symptoms were similar across the two seasons. Conclusions: The clinical presentation of RSV in the two post-COVID seasons suggests that the pathophysiology of the virus did not change across these two years. Further studies are needed to continuously monitor RSV to support an effective prevention strategy.

Keywords: acute respiratory infection; respiratory syncytial virus; bronchiolitis; epidemiology; hypertransaminasemia; surveillance

Citation: Treggiari, D.; Pomari, C.; Zavarise, G.; Piubelli, C.; Formenti, F.; Perandin, F. Characteristics of Respiratory Syncytial Virus Infections in Children in the Post-COVID Seasons: A Northern Italy Hospital Experience. *Viruses* **2024**, *16*, 126. https://doi.org/10.3390/v16010126

Academic Editors: Emanuele Castagno and Irene Raffaldi

Received: 21 November 2023
Revised: 11 January 2024
Accepted: 14 January 2024
Published: 16 January 2024

1. Introduction

Respiratory Syncytial Virus (RSV) is the most common cause of bronchiolitis, leading to hospitalization in infants worldwide and the second cause of infant mortality in resource-limited countries [1,2]. In particular, RSV clinically manifests with bronchiolitis in children younger than two years of age and especially in newborns under six months [3].

Despite the disease burden, no approved vaccines for RSV are currently available. The prevention of RSV infections in infants through maternal vaccination has become a priority and a target for the development of new potential vaccine candidates to be tested in clinical trials [4].

In addition, there are no specific therapeutic options for RSV which is due in part to our limited knowledge of the pathogenesis of the disease [1]. For this reason and the lack of specific etiological treatment, therapy is primarily supportive, based on oxygen and adequate fluid supplementation [5,6].

RSV is distributed worldwide and follows the pattern of influenza; despite strong seasonal trends, the circulation of RSV may increase in periods outside the typical season.

Indeed, the surveillance system in Australia revealed a higher-than-expected number of cases during the austral spring of 2021, from September to December, while winter data indicated a lower-than-expected circulation of RSV [7]. These events were attributed to the impact of the COVID-19 pandemic and related containing measures, which deeply altered the natural course of seasonal viral infections [7–9].

Since the introduction of non-pharmaceutical interventions to control COVID-19, the circulation of RSV in Europe has been limited. Surveillance data from 17 countries showed delayed RSV outbreaks in France (≥12 weeks, w), the UK, and Ireland (≥4 w) during the 2020–21 season [10,11]. Cases of RSV (predominantly affecting young children) in France occurred in older children compared to previous seasons.

In Italy, restrictive measures for the prevention and control of the SARS-CoV-2 pandemic were likely to be responsible for the reduced circulation of RSV and other respiratory tract diseases during the 2020–2021 winter season, as detected by the Respiratory Virus Surveillance Network of the Italian National Institute of Health system (RespiVirNet, https://respivirnet.iss.it/Default.aspx?ReturnUrl=/, accessed on 15 November 2023). Although specific national data on RSV are not yet available, weekly surveillance reports from the Lombardia region confirmed a reduced circulation of the virus as well as a decrease in hospital admissions related to RSV bronchiolitis in that period. SARS-CoV-2 was the only virus detected in samples from individuals with influenza-like illness (ILI) tested between October 2020 and January 2021. These data confirmed the hypothesis that, as for other Countries, in Italy, the SARS-CoV-2 pandemic altered the RSV seasonality. As a result, in autumn 2021, we observed an unexpected surge in RSV infections among infants and elder children compared with previous seasons [12,13].

In this scenario, assays that can detect multiple respiratory viruses play a key role in both treatment decisions and infection control measures. In fact, they allow the monitoring of viruses in real-time as they move through the human population.

In this retrospective study, we evaluated the prevalence and clinical presentation of RSV infection among children referred to our hospital for respiratory problems during the 2021–2022 and 2022–2023 seasons. We compared the two periods, evaluating respiratory virus infection diagnostic and clinical data, including coinfections information, in order to evaluate the evolution of RSV pediatric infection in the post-COVID era.

2. Materials and Methods

2.1. Setting and Participants

This is a single-center retrospective observational study carried out in a hospital in the province of Verona, the IRCCS Hospital Sacro Cuore Don Giovanni Calabria in Negrar di Valpolicella (Verona, Italy). Inclusion criteria were all pediatric patients attending our hospital for suspected respiratory tract infections, with molecular analysis results of influenza virus (either A or B strain, INF-A/B), respiratory syncytial virus (either A or B strain, RSV-A/B), and SARS-CoV-2 test performed from 1 October 2021 to 31 January 2022 (2021–2022 season) and from 15 October 2022 to 31 January 2023 (2022–2023 season). Exclusion criteria were bronchiolitis or other respiratory inflammation with undetermined etiology and subjects transferred to other sites.

2.2. Ethics

The study was conducted according to the guidelines of the Declaration of Helsinki and approved by the Ethical Committee of Verona and Rovigo provinces under the protocol n. 30569 of 22 May 2023.

2.3. Molecular Analysis for SARS-CoV-2, Flu and RSV Infections

Nasopharyngeal swab samples were collected in viral transport media and analyzed with the Anatolia Geneworks Bosphore SARS-CoV-2/Flu/RSV PCR Panel Kit as per the manufacturer's instructions. In total, 510 patients during the 2021–2022 season (from

1 October 2021 to 31 January 2022) and 365 patients during the 2022–2023 season (from 15 October 2022 to 31 January 2023) were subjected to the rt-PCR test.

2.4. Clinical Data Collection

All the patients attended the Pediatric Emergency Department and Operative Unit of Pediatrics of the IRCCS Hospital Sacro Cuore for lower respiratory tract infection symptoms. Clinical data were retrieved from the electronic medical records. The following clinical parameters were considered: length of hospital stay (days), the presence of dyspnea (arterial oxygen saturation $SaO_2 < 95\%$, in addition to respiratory discomfort such as nasal fin flaring, abdominal basculation, intercostal indentation), hypertransaminasemia (alanine aminotransferase, ALT > 40 UI/L), the duration of symptoms (referred to dyspnea, reported in days), aerosol therapy (with salbutamol or budesonide, using the pressurized metered-dose inhaler with valved holding chamber device), and the type of oxygen therapy (low flow if $O_2 < 2$ L/min supplied by nasal prongs; high flow if O_2 is supplied at $\geq$2 L/min by noninvasive oxygen therapy (NIV) using AIRVO Nasal High Flow system). Data were analyzed for each patient using Microsoft Excel 2013 software.

2.5. Statistical Analysis

All variables recorded in this study are presented using descriptive statistics. Continuous variables are presented as the mean ± standard deviation (sd) or median interquartile range (IQR 25th, 75th). Statistical significance was defined for a *p*-value equal to or lesser than 0.05 (Fisher's test, Student's *t*-test). Descriptive and association analyses were conducted using Microsoft Excel supported by the GraphPad statistics program.

3. Results

During the 2021–2022 season, 182 (36%) out of the 510 pediatric subjects (285 males and 225 females) were infected with a respiratory virus, 175 (96%) of whom had RSV-A/B and 7 (4%) had SARS-CoV-2. Seventy-five (41%) patients with lower respiratory tract infections were hospitalized in the pediatrics unit (46 males and 29 females, Table 1), all of them infected by RSV. The median age was 2.1 years.

Table 1. Demographic characteristics of the study population. Population data were divided according to the analyzed seasons. The type and number of detected infections were reported. Lower respiratory tract infection was present in all hospitalized patients. *p* values for the Student's *t*-test were reported.

	Season 2021/2022		Season 2022/2023		
Demographics	**Count**	**Value (%)**	**Count**	**Value (%)**	***p* Value**
Population (n)	510		365		
Gender (n)					
Female	225	44	159	44	
Male	285	56	206	56	
Age (years)					
25% Percentile	1.00		0.70		
Median	2.15		2.60		0.0422
75% Percentile	5.00		6.75		
Infected (n)	182	36	105	29	
RSV	175	96	12	11	
Influenza	0	0	76	72	
SARS-CoV-2	7	4	9	8	
RSV + SARS-CoV-2	0	0	4	4	
RSV + Influenza	0	0	1	1	
RSV + SARS-CoV-2 + Influenza	0	0	2	2	
Hospitalised	75	41	52	49	*p* value

Table 1. *Cont.*

Demographics	Season 2021/2022 Count	Season 2021/2022 Value (%)	Season 2022/2023 Count	Season 2022/2023 Value (%)	*p* Value
Gender (n)					
Female	29	38	20	38	
Male	46	60	32	62	
Age (years)					
25% Percentile	0.1		0.43		
Median	0.4		2.14		0.0001
75% Percentile	1.4		5.43		
Age (years) only for RSV infected					
25% Percentile	0.15		0.06		
Median	0.46		1.10		0.6396
75% Percentile	1.42		2.38		

During the 2022–2023 season, we found 105 (29%) subjects infected with a respiratory virus out of 365 analyzed subjects (206 males and 159 females). The median age for this period was 2.60 years. Differently from the previous season, influenza (72%) was the most accounted for infection, followed by RSV-A/B (11%) and SARS-CoV-2 (8%). Furthermore, during this season, we found different viral coinfections in 7 subjects, i.e., RSV-A/B and SARS-CoV-2 (4 subjects), RSV-A/B and INF-A/B (1 subject), and RSV-A/B, INF-A/B, and SARS-CoV-2 (2 subjects). Coinfections were observed in children <2 years of age. Fifty-two (49%) out of 105 infected subjects were hospitalized for a lower respiratory tract infection (32 males and 20 females, Table 1), among which 100% were RSV-infected patients. Even though it was a minor difference, the median age of the 2021–2022 season was significantly lower (2.15 years, IQR 1.00–5.00) compared to the 2022–2023 season (2.60 years, IQR 0.70–6.753, Fisher's test p = 0.042). This difference increased considering only hospitalized patients, with a median age of 0.46 years (IQR 0.15–1.42) for the first season and 2.14 years (IQR 0.43–5.43) for the second season (p = 0.0001). Figure 1 shows the trend of RSV-A/B, INF-A/B, and SARS-CoV-2-positive children detected during the considered periods.

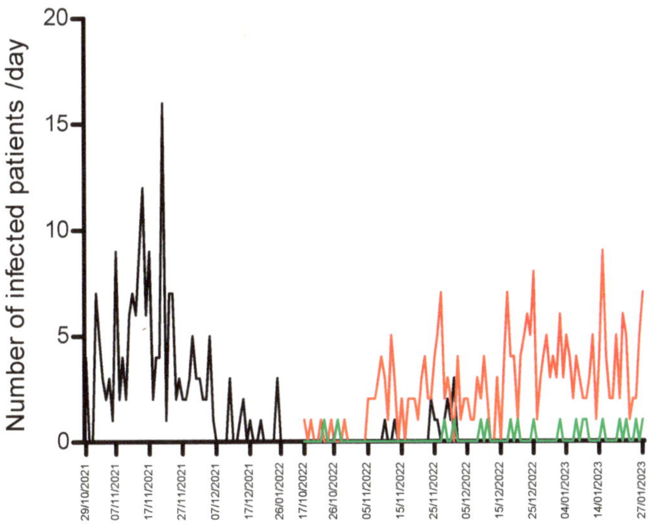

Figure 1. Epidemic trend of RSV, influenza, and SARS-CoV-2 infections from hospitalized children in our hospital during the seasons 2021–2022 and 2022–2023. The lines are represented by the following color code: Black line for RSV; red line for influenza; green line for SARS-CoV-2.

Focusing on the 2021–2022 season, the number of infected children increased gradually from October 2021 and reached a peak of 14 positive diagnoses per day around mid-November 2021. After this date, the numbers slowly declined, with less conspicuous peaks. From the end of the month, the decline was even more prominent, leading to few cases being detected in December. No additional new cases were detected in January 2022. Conversely, the RSV-A/B trend was much flatter during the 2022–2023 season, in which the maximum number of RSV-A/B positive cases was detected only at the beginning of December 2022 (4 cases per day), while the most prevalent infection was represented by INF-A/B, starting from October 2022 and spreading all over the season. All the positive flu cases were sent to the regional referral center for genetic characterization, and the most circulated influenza genotype was identified as A/H_3N_2.

We further analyzed the clinical characteristics of the hospitalized patients, and the data are summarized in Table 2. The following parameters were considered: the length of hospitalization, the presence of dyspnea, the presence of a high level of alanine aminotransferase (ALT > 40 UI/L), the duration of symptoms, and the type of oxygen therapy (low flow $O_2 < 2$ L/min, or high flow with $O_2 \geq 2$ L/min administered by AIRVO).

Table 2. Clinical characteristics of hospitalized patients. Data on hospital length and duration of symptoms were reported as mean ± standard deviation (sd) in days. Student's *t*-test and Fisher's test *p*-values were also reported, considering season 2022–2023 versus season 2021–2022 data—the duration of symptoms referred to the presence of dyspnea.

	Season 2021/2022		**Season 2022/2023**		
	Count	**Value (%)**	**Count**	**Value (%)**	***p*-Value**
Number of patients (n)	75	41	52	46	
Infections (lower respiratory tract), (n)					
RSV	75	100	12	23	
Influenza	0	0	28	53	
SARS-CoV-2	0	0	5	10	
Co-infections	0	0	7	14	
Hospital lenght of stay (mean ± sd, days)					
Overall	5.2 ± 2.7	-	5.0 ± 2.5	-	ns
RSV	5.2 ± 2.7	-	5.0 ± 2.5	-	ns
Influenza	0	-	4.7 ± 2.5	-	
SARS-CoV-2	0	-	4.0 ± 2.5	-	
RSV + SARS-CoV-2	0	-	6.5 ± 3.4	-	
RSV + Influenza	0	-	4	-	
RSV + SARS-CoV-2 + Influenza	0	-	6	-	
Clinical characteristics					
Dyspnea (n)	52	69	21	40	0.0034
ALT (>40 UI/L), (n)	7	9	4	7	ns
Duration of symptoms (mean ± sd, days)	9.5 ± 0.4	-	7.8 ± 0.6	-	0.0255
Therapeutic interventions (n)					
Areosol	73	97	32	62	<0.0001
O_2 (<2 L/min)	41	56	16	31	0.0063
AIRVO (≥2 L/min)	22	29	5	10	0.0044
Steroids	68	90	31	60	<0.0001

We did not find any statistically significant difference in the average length of the hospital stay during the two seasons (mean ± sd, 5.0 ± 2.7 vs. 5.2 ± 2.7 days), even when considering only RSV-A/B inpatients. Moreover, we found that neither the type of virus nor the presence of coinfections had an impact on the duration of hospitalization during the 2022–2023 season (Table 2). Our data showed that dyspnea was present in 69% of cases in the first season; conversely, in the following season, only 40% of subjects showed this symptom ($p = 0.0034$). A significative difference was also found in the duration of symptomatology between the two periods, with a longer presence of symptoms observed in the 2021–2022 season compared to the 2022–2023 season (mean diff ± sd, 1.7 ± 0.7 days, $p = 0.0255$). When comparing the two seasons solely for RSV-A/B, we did not find significant differences in terms of both dyspnea (51 out of 75 during 2021/2022, 9 out of 12 during 2022/2023, $p = 1.0$, Fisher's test) and symptoms duration (difference between means 1.2 ± 1.1, $p = 0.307$). No difference in terms of the presence of hypertransaminasemia was found between the two seasons ($p = 0.713$).

When analyzing the therapeutic interventions between the two seasons, we found statistically significant differences in aerosol ($p < 0.0001$), oxygen therapy with low flow ($p = 0.0063$), and AIRVO ($p = 0.0044$, Table 2), with a larger requirement of all the three therapies during the 2021–2022 season. For both seasons, almost all RSV children were given aerosol therapy (97 and 100% for the 2021–2022 and 2022–2023 periods, respectively), often because this therapy had already been started at home by the caregiver. Steroids are not indicated in pediatric guidelines, but in clinical practice, steroids are given to bronchiolitis patients for prompt improvement, favoring a positive recovery. No differences in steroid treatment were observed for RSV patients among the two seasons [14]. In the 2021–2022 season, 41 out of the 75 RSV inpatients required oxygen therapy (56%). Among patients receiving oxygen, twenty-two (53%) required the use of NIV at high flows with the aid of the AIRVO system for severe dyspnea. The mean duration of used high-flow oxygen therapy was 6 ± 2.4 days. The 22 subjects who needed high flow (29% of the RSV-A/B patients) were patients with a lower median age than subjects who needed low flow or no oxygen support. In addition, two of them were former preterm patients of 36 weeks (w) and 35 w ± 5 days. The latter subjects had a longer duration of high-flow respiratory therapy (12 and 13 days, respectively), with a longer duration of symptoms (19 days) and consequently longer hospitalization (12 and 13 days, respectively). In the 2022–2023 season, 31% of patients required low-flow oxygen therapy, and 10% needed AIRVO support. For both seasons, patients who needed high-flow oxygen therapy had a lower median age compared to all the hospitalized patients in the respective periods: 0.18 years (IQR 0.08–0.53) for the first season (−0.28 years compared to the median age of hospitalized patients in 2021–2022) and 1.94 years (IQR 1.72–7.80) for the second season (−0.20 years compared to the hospitalized patients in 2022–2023). This supports the hypothesis that younger babies are more subject to severe symptoms. Considering only the NIV-treated patients, the two seasons were statistically different ($p = 0.004$, Fisher test), but these data probably reflect the difference in the age of the hospitalized patients in the two seasons. However, no significant differences in terms of therapeutic interventions were found in RSV patients between the two seasons, with 58% of the 2022–2023 RSV patients requiring low flow O_2 and 58% AIRVO. Ninety-nine percent (99%) of the reported RSV hospitalized cases manifested bronchiolitis, and only 1% manifested broncho-pneumonia, specifically a subject with comorbidity (pulmonary valve stenosis).

4. Discussion

RSV is one of the most common respiratory viruses. It affects not only young children but also the elderly and immunocompromised patients [15,16]. With the emergence of SARS-CoV-2, a considerable decrease in RSV incidence and hospitalization rate was observed worldwide during the 2020–2021 cold season [17–22], coinciding with the implementation of public health and social containment measures. A seasonality shift and a delayed RSV outbreak with a greater number of infected patients were reported in the 2021–2022 season in several countries, such as Australia, Saudi Arabia, New Zealand, France, the UK, and Japan [23–28]. Also, in Italy, an extraordinary surge in RSV was observed in the fall of 2021 [12,13,29], with hospitalization rates similar to the previous years but with a higher rate of admission to intensive care units [30]. Among others, the waning immunity against RSV in children and adults and a lack of antibody transmission from the mothers to the newborns seem to be the primary factors responsible for the exceptional resurgence of RSV in the 2021–2022 season [30]. This has led to increased attention towards RSV infections, which must be monitored to evaluate the evolving epidemiology and clinical manifestations. In our study, we assessed the clinical characteristics, in terms of symptoms and the subsequent healthcare management, of RSV pediatric infections in children admitted to our hospital one and two years after the relaxation of containment measures implemented during the COVID-19 pandemic.

In the first analyzed season (2021–2022), the number of RSV-affected children was very high, in line with literature data, especially among children under one year of age [31]. The outbreak appeared in a shorter interval, between October and December 2021, anticipating the expected epidemic by a couple of months and making it shorter. Pre-pandemic levels of infections were overcome, and, as a result, there was an overload in our hospital emergency rooms and the whole country [31]. A relevant feature of the 2021–2022 season was the severity of bronchiolitis: a more severe clinical presentation and a frequent need for high-flow oxygen therapy, especially in younger children compared to the previous years. In fact, in previous years, during RSV outbreaks in our hospital, it had never been necessary to use AIRVO for high-flow oxygen therapy and for such a prolonged time. There were significantly fewer cases of hospitalized children (lower than 30/years), with mostly home management. Unfortunately, because molecular testing was not often required in the pre-COVID-19 era and diagnosis was mostly clinical, we do not have sufficient data today to compare with what was observed in the post-COVID-19 era [29,31].

In the second analyzed season (2022–2023), a reduced number of infected pediatric patients were detected compared to the first post-COVID year (28% vs. 36%) but with a similar hospitalization rate (46% vs. 40%). During the second season, different viral infections were registered, with a predominance of the influenza virus (72%), while the RSV accounted for 12% of the detected infections. Another aspect that emerged in our study is the total absence of coinfections in the fall 2021 and their appearance in the fall 2022 season (7% of the total infected patients, 14% of the hospitalized). Coinfections were observed in children <2 years of age, suggesting that respiratory coinfections in children with SARS-CoV-2 are common, above all in younger children. This could be due to an immature immune response. A recent paper highlighted the role of interferons in preventing coinfections by different viruses [32]. In neonates and young children, the pathways involved in interferon production are still in a developmental state and can be less protective [33]. Chuang et al., reviewing the post-COVID pediatric infections data, reported a similar trend of SARS-CoV-2 and RSV coinfections in several regions [34]. The clinical effect of SARS-CoV-2 and RSV coinfection is still a debated issue; in fact, it seems to be correlated with a longer hospital stay, although a direct correlation with an increased mortality or intensive care unit admission has not been demonstrated [34,35].

Our study highlighted a decrease in the severity of symptoms among hospitalized pediatric patients in the second post-COVID year. Nevertheless, focusing only on RSV patients, a similar level of dyspnea, duration of symptoms, and need for oxygen therapy was observed, indicating that RSV clinical presentation was not changed across the two different seasons. Thus, the less severe symptoms registered globally in the 2022–2023 hospitalized population of pediatric patients were due to the presence of different viruses, such as SARS-CoV-2 and INFA/B.

The present study has the following limitations that should be considered. First, it was a retrospective analysis within a single hospital; testing was performed at the clinician's discretion; clinical data were retrieved retrospectively, reading the letter of discharge from hospitalization; the data from the pre-COVID-19 era are missing. Moreover, we did not characterize the circulating RSV strains at the genomic level to evaluate possible differences between the two periods.

5. Conclusions

In this retrospective study, we pointed out that in the two years after the relaxation of the social containment measures, we observed a huge "pure" RSV outbreak in the 2021–2022 season, in which 96% of infected pediatric subjects carried RSV, followed by a more "mixed" 2022–2023 season, in which flu was prevalent, and RSV infections were reduced to 12%, including some coinfections. The clinical presentations of RSV in the two seasons were similar, suggesting that the pathophysiology of the virus has not changed across these two years. Further studies are needed to continuously monitor this virus to support the creation of an effective year-round RSV-specific prevention strategy and monitor the presence of respiratory viruses regardless of their seasonality.

Author Contributions: Conceptualization, F.P., C.P. (Chiara Piubelli) and G.Z.; investigation, C.P. (Chiara Pomari) and D.T.; formal analysis, data curation, C.P. (Chiara Pomari) and D.T.; writing-original draft preparation, D.T.; writing-review and editing, C.P. (Chiara Piubelli); supervision, F.P., C.P. (Chiara Piubelli) and F.F. All authors have read and agreed to the published version of the manuscript.

Funding: This research was funded by the Italian Ministry of Health "Fondi Ricerca Corrente, Project L1P6" for IRCCS Sacro Cuore–Don Calabria Hospital and by EU funding within the MUR PNRR Extended Partnership initiative on Emerging Infectious Diseases (Project no. PE00000007, INF-ACT).

Institutional Review Board Statement: The study was conducted according to the guidelines of the Declaration of Helsinki and approved by the Ethical Committee of Verona and Rovigo provinces under the protocol n. 30569 of 22 May 2023.

Informed Consent Statement: Being a retrospective study, any reasonable effort was applied to obtain informed consent for data use, but this was not possible in the vast majority of cases. Therefore, in compliance with Art 21 of the Italian "decreto legislative" 10 August 2018 n. 101, formal approval from the competent Ethics Committee was obtained.

Data Availability Statement: The data supporting this study's findings are available in the Zenodo Repository at https://doi.org/10.5281/zenodo.10159164 (accessed on 20 November 2023).

Conflicts of Interest: The authors declare no conflicts of interest, and the funders had no role in the design of the study, in the collection, analyses, or interpretation of data, in the writing of the manuscript, or in the decision to publish the results.

References

1. Brenes-Chacon, H.; Garcia-Mauriño, C.; Moore-Clingenpeel, M.; Mertz, S.; Ye, F.; Cohen, D.M.; Ramilo, O.; Mejias, A. Age-Dependent Interactions Among Clinical Characteristics, Viral Loads and Disease Severity in Young Children With Respiratory Syncytial Virus Infection. *Pediatr. Infect. Dis. J.* **2021**, *40*, 116–122. [CrossRef]
2. Hall, C.B.; Weinberg, G.A.; Iwane, M.K.; Blumkin, A.K.; Edwards, K.M.; Staat, M.A.; Auinger, P.; Griffin, M.R.; Poehling, K.A.; Erdman, D.; et al. The Burden of Respiratory Syncytial Virus Infection in Young Children. *N. Engl. J. Med.* **2009**, *360*, 588–598. [CrossRef] [PubMed]
3. Hirve, S.; Crawford, N.; Palekar, R.; Zhang, W.; WHO RSV surveillance Group. Clinical Characteristics, Predictors, and Performance of Case Definition—Interim Results from the WHO Global Respiratory Syncytial Virus Surveillance Pilot. *Influenza Resp. Viruses* **2020**, *14*, 647–657. [CrossRef]
4. Abu-Raya, B.; Maertens, K.; Edwards, K.M.; Omer, S.B.; Englund, J.A.; Flanagan, K.L.; Snape, M.D.; Amirthalingam, G.; Leuridan, E.; Damme, P.V.; et al. Global Perspectives on Immunization During Pregnancy and Priorities for Future Research and Development: An International Consensus Statement. *Front. Immunol.* **2020**, *11*, 1282. [CrossRef] [PubMed]
5. Azzari, C.; Baraldi, E.; Bonanni, P.; Bozzola, E.; Coscia, A.; Lanari, M.; Manzoni, P.; Mazzone, T.; Sandri, F.; Checcucci Lisi, G.; et al. Epidemiology and Prevention of Respiratory Syncytial Virus Infections in Children in Italy. *Ital. J. Pediatr.* **2021**, *47*, 198. [CrossRef]
6. Biagi, C.; Scarpini, S.; Paleari, C.; Fabi, M.; Dondi, A.; Gabrielli, L.; Gennari, M.; Lanari, M.; Pierantoni, L. Impact of Guidelines Publication on Acute Bronchiolitis Management: 10-Year Experience from a Tertiary Care Center in Italy. *Microorganisms* **2021**, *9*, 2221. [CrossRef] [PubMed]
7. Foley, D.A.; Yeoh, D.K.; Minney-Smith, C.A.; Martin, A.C.; Mace, A.O.; Sikazwe, C.T.; Le, H.; Levy, A.; Moore, H.C.; Blyth, C.C. The Interseasonal Resurgence of Respiratory Syncytial Virus in Australian Children Following the Reduction of Coronavirus Disease 2019–Related Public Health Measures. *Clin. Infect. Dis.* **2021**, *73*, e2829–e2830. [CrossRef]

8. Manti, S.; Giallongo, A.; Parisi, G.F.; Papale, M.; Presti, S.; Lo Bianco, M.; Spicuzza, L.; Leonardi, S. Impact of COVID-19 Pandemic and Lockdown on the Epidemiology of RSV-Mediated Bronchiolitis: Experience from Our Centre. *Children* **2022**, *9*, 1723. [CrossRef]
9. Stera, G.; Pierantoni, L.; Masetti, R.; Leardini, D.; Biagi, C.; Buonsenso, D.; Pession, A.; Lanari, M. Impact of SARS-CoV-2 Pandemic on Bronchiolitis Hospitalizations: The Experience of an Italian Tertiary Center. *Children* **2021**, *8*, 556. [CrossRef]
10. Roland, D.; Williams, T.; Lyttle, M.D.; Marlow, R.; Hardelid, P.; Sinha, I.; Swann, O.; Maxwell-Hodkinson, A.; Cunningham, S. Features of the Transposed Seasonality of the 2021 RSV Epidemic in the UK and Ireland: Analysis of the First 10,000 Patients. *Arch. Dis. Child* **2022**, *107*, 1062–1063. [CrossRef]
11. Vaux, S.; Viriot, D.; Forgeot, C.; Pontais, I.; Savitch, Y.; Barondeau-Leuret, A.; Smadja, S.; Valette, M.; Enouf, V.; Parent Du Chatelet, I. Bronchiolitis Epidemics in France during the SARS-CoV-2 Pandemic: The 2020–2021 and 2021–2022 Seasons. *Infect. Dis. Now.* **2022**, *52*, 374–378. [CrossRef]
12. Bozzola, E. Respiratory Syncytial Virus Resurgence in Italy: The Need to Protect All Neonates and Young Infants. *Ijerph* **2021**, *19*, 380. [CrossRef] [PubMed]
13. Treggiari, D.; Piubelli, C.; Caldrer, S.; Mistretta, M.; Ragusa, A.; Orza, P.; Pajola, B.; Piccoli, D.; Conti, A.; Lorenzi, C.; et al. SARS-CoV-2 Rapid Antigen Test in Comparison to RT-PCR Targeting Different Genes: A Real-life Evaluation among Unselected Patients in a Regional Hospital of Italy. *J. Med. Virol.* **2022**, *94*, 1190–1195. [CrossRef] [PubMed]
14. Fernandes, R.M.; Hartling, L. Glucocorticoids for Acute Viral Bronchiolitis in Infants and Young Children. *JAMA* **2014**, *311*, 87. [CrossRef] [PubMed]
15. Bouzid, D.; Visseaux, B.; Ferré, V.M.; Peiffer-Smadja, N.; Le Hingrat, Q.; Loubet, P. Respiratory Syncytial Virus in Adults with Comorbidities: An Update on Epidemiology, Vaccines, and Treatments. *Clin. Microbiol. Infect.* **2023**, *29*, S1198743X23004123. [CrossRef]
16. Staadegaard, L.; Caini, S.; Wangchuk, S.; Thapa, B.; De Almeida, W.A.F.; De Carvalho, F.C.; Njouom, R.; Fasce, R.A.; Bustos, P.; Kyncl, J.; et al. The Global Epidemiology of RSV in Community and Hospitalized Care: Findings From 15 Countries. *Open Forum Infect. Dis.* **2021**, *8*, ofab159. [CrossRef]
17. Curatola, A.; Lazzareschi, I.; Bersani, G.; Covino, M.; Gatto, A.; Chiaretti, A. Impact of COVID-19 Outbreak in Acute Bronchiolitis: Lesson from a Tertiary Italian Emergency Department. *Pediatr. Pulmonol.* **2021**, *56*, 2484–2488. [CrossRef]
18. Kadambari, S.; Goldacre, R.; Morris, E.; Goldacre, M.J.; Pollard, A.J. Indirect Effects of the COVID-19 Pandemic on Childhood Infection in England: Population Based Observational Study. *BMJ* **2022**, *376*, e067519. [CrossRef]
19. Li, Y.; Wang, X.; Blau, D.M.; Caballero, M.T.; Feikin, D.R.; Gill, C.J.; Madhi, S.A.; Omer, S.B.; Simões, E.A.F.; Campbell, H.; et al. Global, Regional, and National Disease Burden Estimates of Acute Lower Respiratory Infections Due to Respiratory Syncytial Virus in Children Younger than 5 Years in 2019: A Systematic Analysis. *Lancet* **2022**, *399*, 2047–2064. [CrossRef]
20. Trenholme, A.; Webb, R.; Lawrence, S.; Arrol, S.; Taylor, S.; Ameratunga, S.; Byrnes, C.A. COVID-19 and Infant Hospitalizations for Seasonal Respiratory Virus Infections, New Zealand, 2020. *Emerg. Infect. Dis.* **2021**, *27*, 641–643. [CrossRef]
21. Willis, Z.; De St Maurice, A. COVID-19 Control Measures and Common Paediatric Infections. *BMJ* **2022**, *376*, n3093. [CrossRef] [PubMed]
22. Yeoh, D.K.; Foley, D.A.; Minney-Smith, C.A.; Martin, A.C.; Mace, A.O.; Sikazwe, C.T.; Le, H.; Levy, A.; Blyth, C.C.; Moore, H.C. Impact of Coronavirus Disease 2019 Public Health Measures on Detections of Influenza and Respiratory Syncytial Virus in Children During the 2020 Australian Winter. *Clin. Infect. Dis.* **2021**, *72*, 2199–2202. [CrossRef]
23. Alkharsah, K.R. The Scope of Respiratory Syncytial Virus Infection in a Tertiary Hospital in the Eastern Province of Saudi Arabia and the Change in Seasonal Pattern during and after the COVID-19 Pandemic. *Medicina* **2022**, *58*, 1623. [CrossRef] [PubMed]
24. Lumley, S.F.; Richens, N.; Lees, E.; Cregan, J.; Kalimeris, E.; Oakley, S.; Morgan, M.; Segal, S.; Dawson, M.; Walker, A.S.; et al. Changes in Paediatric Respiratory Infections at a UK Teaching Hospital 2016–2021; Impact of the SARS-CoV-2 Pandemic. *J. Infect.* **2022**, *84*, 40–47. [CrossRef] [PubMed]
25. Huang, Q.S.; Wood, T.; Jelley, L.; Jennings, T.; Jefferies, S.; Daniells, K.; Nesdale, A.; Dowell, T.; Turner, N.; Campbell-Stokes, P.; et al. Impact of the COVID-19 Nonpharmaceutical Interventions on Influenza and Other Respiratory Viral Infections in New Zealand. *Nat Commun* **2021**, *12*, 1001. [CrossRef]
26. Marriott, D.; Beresford, R.; Mirdad, F.; Stark, D.; Glanville, A.; Chapman, S.; Harkness, J.; Dore, G.J.; Andresen, D.; Matthews, G.V. Concomitant Marked Decline in Prevalence of Severe Acute Respiratory Syndrome Coronavirus 2 (SARS-CoV-2) and Other Respiratory Viruses Among Symptomatic Patients Following Public Health Interventions in Australia: Data from St Vincent's Hospital and Associated Screening Clinics, Sydney, NSW. *Clin. Infect. Dis.* **2021**, *72*, e649–e651. [CrossRef] [PubMed]
27. Takashita, E.; Kawakami, C.; Momoki, T.; Saikusa, M.; Shimizu, K.; Ozawa, H.; Kumazaki, M.; Usuku, S.; Tanaka, N.; Okubo, I.; et al. Increased Risk of Rhinovirus Infection in Children during the Coronavirus Disease-19 Pandemic. *Influenza Resp Viruses* **2021**, *15*, 488–494. [CrossRef] [PubMed]
28. Delestrain, C.; Danis, K.; Hau, I.; Behillil, S.; Billard, M.; Krajten, L.; Cohen, R.; Bont, L.; Epaud, R. Impact of COVID-19 Social Distancing on Viral Infection in France: A Delayed Outbreak of RSV. *Pediatr. Pulmonol.* **2021**, *56*, 3669–3673. [CrossRef]
29. Loconsole, D.; Centrone, F.; Rizzo, C.; Caselli, D.; Orlandi, A.; Cardinale, F.; Serio, C.; Giordano, P.; Lassandro, G.; Milella, L.; et al. Out-of-Season Epidemic of Respiratory Syncytial Virus during the COVID-19 Pandemic: The High Burden of Child Hospitalization in an Academic Hospital in Southern Italy in 2021. *Children* **2022**, *9*, 848. [CrossRef]

30. Koltai, M.; Krauer, F.; Hodgson, D.; Van Leeuwen, E.; Treskova-Schwarzbach, M.; Jit, M.; Flasche, S. Determinants of RSV Epidemiology Following Suppression through Pandemic Contact Restrictions. *Epidemics* **2022**, *40*, 100614. [CrossRef]
31. Faraguna, M.C.; Lepri, I.; Clavenna, A.; Bonati, M.; Vimercati, C.; Sala, D.; Cattoni, A.; Melzi, M.L.; Biondi, A. The Bronchiolitis Epidemic in 2021–2022 during the SARS-CoV-2 Pandemic: Experience of a Third Level Centre in Northern Italy. *Ital. J. Pediatr.* **2023**, *49*, 26. [CrossRef] [PubMed]
32. Cohen, J. Will Viral Interference Hold off the Tripledemic? *Science* **2022**, *378*, 814–815. [CrossRef] [PubMed]
33. Maddux, A.B.; Douglas, I.S. Is the Developmentally Immature Immune Response in Paediatric Sepsis a Recapitulation of Immune Tolerance? *Immunology* **2015**, *145*, 1–10. [CrossRef] [PubMed]
34. Chuang, Y.-C.; Lin, K.-P.; Wang, L.-A.; Yeh, T.-K.; Liu, P.-Y. The Impact of the COVID-19 Pandemic on Respiratory Syncytial Virus Infection: A Narrative Review. *IDR* **2023**, *16*, 661–675. [CrossRef]
35. Alvares, P.A. SARS-CoV-2 and Respiratory Syncytial Virus Coinfection in Hospitalized Pediatric Patients. *Pediatr. Infect. Dis. J.* **2021**, *40*, e164–e166. [CrossRef]

viruses

MDPI

Brief Report

Urgent Hospitalizations Related to Viral Respiratory Disease in Children during Autumn and Winter Seasons 2022/2023

Francesca Peri [1,*], Beatrice Lorenzon [1], Carolina Cason [2], Alessandro Amaddeo [3], Stefania Norbedo [3], Manola Comar [2,4], Egidio Barbi [1,5] and Giorgio Cozzi [3]

1 Department of Medicine, Surgery and Health Sciences, University of Trieste, 34100 Trieste, Italy; beatricelorenzon@gmail.com (B.L.); egidio.barbi@burlo.trieste.it (E.B.)
2 SSD of Advanced Microbiology Diagnosis and Translational Research, Institute for Maternal and Child Health-IRCCS Burlo Garofolo, 34100 Trieste, Italy; carolina.cason@burlo.trieste.it (C.C.); manola.comar@burlo.trieste.it (M.C.)
3 Emergency Department, Institute for Maternal and Child Health-IRCCS Burlo Garofolo, 34100 Trieste, Italy; alessandro.amaddeo@burlo.trieste.it (A.A.); stefania.norbedo@burlo.trieste.it (S.N.); giorgio.cozzi@burlo.trieste.it (G.C.)
4 Department of Medical Sciences, University of Trieste, Piazzale Europa 1, 34127 Trieste, Italy
5 Department of Pediatrics, Institute for Maternal and Child Health IRCCS "Burlo Garofolo", 34100 Trieste, Italy
* Correspondence: francesca.peri@burlo.trieste.it

Abstract: Aim: The loosening of social distancing measures over the past two years has led to a resurgence of seasonal epidemics associated with respiratory viral infections in children. We aim to describe the impact of such infections through urgent hospitalizations in a pediatric emergency department. Methods: We performed a retrospective review of medical records of all children and adolescents with a positive nasal swab admitted at the children's hospital IRCCS Burlo Garofolo of Trieste, in Italy, from September 2021 to March 2022, and September 2022 to March 2023. Results: Respiratory Syncytial Virus and Influenza viruses accounted for up to 55% of hospitalizations for respiratory infections during the study periods. During the last season, the number of hospitalizations related to the Influenza virus was five times higher than those related to SARS-CoV-2 (25% vs. 5%). Respiratory Syncytial Virus was associated with a greater need for respiratory support, mostly HFNC (High Flow Nasal Cannula). Conclusions: Respiratory Syncytial Virus and Influenza virus had a more significant impact on urgent hospitalizations during the past wintery seasons than SARS-CoV-2.

Keywords: Respiratory Syncytial Virus (RSV); bronchiolitis; emergency department

Citation: Peri, F.; Lorenzon, B.; Cason, C.; Amaddeo, A.; Norbedo, S.; Comar, M.; Barbi, E.; Cozzi, G. Urgent Hospitalizations Related to Viral Respiratory Disease in Children during Autumn and Winter Seasons 2022/2023. *Viruses* **2023**, *15*, 2425. https://doi.org/10.3390/v15122425

Academic Editors: Emanuele Castagno and Irene Raffaldi

Received: 16 November 2023
Revised: 7 December 2023
Accepted: 13 December 2023
Published: 14 December 2023

1. Introduction

Since 2020, the introduction of social distancing measures limiting the spread of the COronaVIrus Disease 19 (COVID-19) pandemic has wholly changed the epidemiology of common viral infections in children. These measures included a national lockdown with school closures from March to May 2020, followed by the mandatory use of face masks and different levels of social distancing measures from November 2020 to May 2021. These measures have had a profound impact on the transmission of viruses through respiratory droplets [1]. During the first winter of the COVID-19 pandemic, the disappearance of winter epidemics sustained by Respiratory Syncytial Virus (RSV) and Influenza virus was observed. In contrast, the prevalence of Rhinoviruses remained largely unaffected, likely due to their viral stability and the role of asymptomatic transmission [2–5]. In the next two years, the progressive loosening of most social distancing measures led to a resurgence of seasonal epidemics associated with respiratory viral infections in children. Initially, several European countries reported out-of-season RSV outbreaks during the spring of 2021 [6,7]. At the same time, an RSV epidemic was observed during the autumn of 2021, showing an increase in disease severity [7–9]. Finally, the first Influenza virus epidemic was reported in the autumn of 2022, along with RSV and the COVID-19 pandemic [10]. The burden on urgent hospitalizations related to the unprecedented combination of three highly diffusible and virulent infections has not yet been described. This report aimed to describe the impact of different respiratory viruses on urgent hospital admissions in the north-east of Italy, with particular focus on RSV, Influenza virus and SARS-CoV-2 during the autumn–winter season in 2022/23 compared to the same period in the year before.

2. Methods

We conducted a retrospective study reviewing the medical records of all children and adolescents hospitalized at the pediatric emergency department (PED) at the tertiary level, in the University Teaching Children's Hospital, Institute for Maternal and Child Health, IRCCS Burlo Garofolo of Trieste, Italy, from 1 September 2021 to 31 March 2022 and from 1 September 2022 to 31 March 2023. Since the summer of 2021, all children attending the PED with fever or respiratory symptoms and who needed hospitalization underwent a multiple nucleic acid amplification assay using a nasopharyngeal (NP) swab for 13 common viral respiratory agents, including RSV, SARS-CoV-2, Influenza virus and Rhinovirus (Respiratory Flow Chip assay—Vitro, Sevilla, Spain).

Eligible patients were children and adolescents, from 0 to 17 years of age, who were admitted and hospitalized for at least 24 h at the PED. Among eligible patients, all cases that tested positive for a viral agent at the swab test were retained for analysis. For every selected patient, the following data were recorded: age, gender, diagnosis and the need for ventilatory support. The triage code at arrival was also registered. Italian pediatric triage codes consist of a five grade scale; for the purpose of our study, "urgent code" was the highest clinical priority code. The number of patients who arrived at the PED during the two study periods and the number of viral swabs performed were also collected.

The primary study outcome was the number of patients hospitalized for RSV, Influenza virus or SARS-CoV-2 infection between the 2021/2022 and 2022/2023 autumn and winter seasons.

2.1. Ethics

The Institutional Review Board (IRB) of the Institute gave ethical approval to the study protocol (RC 10/2020). Due to the study's retrospective nature, no specific written informed consent was administered.

2.2. Statistical Analysis

Data of enrolled children were summarized using descriptive analysis. Categorical variables were reported through absolute frequencies and percentages.

We divided the study cohort into two periods according to the patient's arrival date: 1 September 2021–31 March 2022 and 1 September 2022–31 March 2023.

The chi-square and Fisher's Exact tests were used to examine the variables of interest in the differences between the two periods. Data were entered into an Excel spreadsheet, and statistical analyses were performed with R software (version 4.0.3, 2020). Statistical significance was considered for p-values < 0.05.

3. Results

During the study period, 30.718 PED visits were performed: 13.706 (44.6%) in autumn–winter 2021/22 and 17.012 (55.4%) in autumn–winter 2022/23. Of these, 97 (0.7%) and 188 patients (1.1%) with a positive multi-viral NP swab were hospitalized during autumn–winter 2021/22 and 2022/23, respectively. Forty-six patients (47%) were female in the first group, and eighty-two (44%) in the second. The mean age was significantly lower in the first period (2.0 years $\pm$ 3.3) compared to the second period (3.1 years $\pm$ 4.3) (p = 0.028). Table 1 summarizes the impact of RSV, Influenza virus and SARS-CoV-2 on urgent hospitalizations during the autumn–winter season 2022/23, compared to the same period the year before. During autumn and winter 2021/2022, three urgent hospitalizations related to Influenza virus infection were recorded. On the other hand, the number of urgent hospitalizations related to the Influenza virus rose to five times higher than those related to SARS-CoV-2 over the last winter season, 48 (25%) vs. 10 (5%). RSV and Influenza viruses accounted for up to 55% of hospitalizations for respiratory infections. In our population, RSV accounted for most urgent hospitalizations in the autumn–winter season 2021/2022, 27 (28%) and, 57 (30%) in the following season. Rhinovirus was the most prevalent in our population in both seasons: 38 (39%) in 2021/22 and 75 (40%) in 2022/23. It played a dual role as a cause of both coinfections and primary infections, accounting for 31 cases of isolated infection in 2021/2022 and 11 cases in 2022/2023. During autumn–winter 2021/2022, six cases (19%) were associated with bronchiolitis and four (12%) required respiratory support with HFNC. In the subsequent season, two cases (18%) resulted in bronchiolitis and only one (9%) required HFNC support. No patients in either season required support with CPAP.

Table 1. Virological data for urgent hospitalizations.

<table>
<tr><th></th><th>2021/22</th><th>2022/23</th></tr>
<tr><td>Number of PED visits</td><td>13.706</td><td>17.012</td></tr>
<tr><td>Number of urgent hospitalizations</td><td>110</td><td>226</td></tr>
<tr><td rowspan="2">Number of urgent hospitalizations with positive NP swab</td><td>N = 97 *
RSV = 27 (28%)
Influenza = 3 (3%)
SARS-CoV-2 = 22 (23%)</td><td>N = 188 *
RSV = 57 (30%)
Influenza = 48 (25%)
SARS-CoV-2 = 10 (5%)</td></tr>
<tr><td>Others:
• Rhinovirus = 38 (39%)
• Parainfluenza virus = 3 (3%)
• Adenovirus = 8 (8%)
• B. Pertussis = 0</td><td>Others:
• Rhinovirus = 75 (40%)
• Parainfluenza virus = 7 (4%)
• Adenovirus = 14 (7%)</td></tr>
<tr><td>Co-infections</td><td>• RSV + rhinovirus = 5 (5%)
• RSV + SARS-CoV-2 = 1 (1%)
• RSV + other coronaviruses = 2 (2%)
• SARS-CoV-2 + rhinovirus = 1 (1%)</td><td>• RSV + rhinovirus = 15 (8%)
• RSV + SARS-CoV-2 = 1 (0.5%)
• RSV + influenza = 4 (2%)
• SARS-CoV-2 + influenza = 0
• Influenza + rhinovirus = 11 (6%)
• SARS-CoV-2 + rhinovirus = 0</td></tr>
</table>

List of abbreviations: pediatric emergency department = PED, nasopharyngeal = NP, respiratory syncytial virus = RSV, severe acute respiratory syndrome coronavirus 2 = SARS-CoV-2. * The total number refers to the number of positive tests; the sum of individual viruses is higher due to co-infections.

The few cases of coinfections did not allow a dedicated subgroup analysis. Nevertheless, we did not record a greater severity in those patients with coinfections.

Table 2 summarizes clinical data according to virus group, comparing the impact of each virus on the urgent triage code and the need for respiratory support between the two years. In our cohort, RSV was responsible for most of the urgent triage codes during the autumn–winter seasons of 2021/2022 and 2022/2023: 14 (38%) and 30 (49%), respectively. However, it must be highlighted that Rhinovirus was responsible for urgent triage codes in 31 (84%) cases during the autumn–winter season 2021/2022. It represents the cause of 67% of bronchiolitis during both 2021 and 2022 and was associated with greater need for respiratory support, mostly HFNC (7, 26%, during 2021 and 21, 37%, during 2022). No statistical differences were found between the two groups.

Table 2. Demographic characteristics of the population according to virological tests.

	RSV			Influenza			SARS-CoV-2			Rhinoviruses	
	2021/2022	2022/2023	p	2021/2022	2022/2023	p	2021/2022	2022/2023	p	2021/2022 2022/2023	p
Age, years (mean, SD)	1.2 ± 2.1	1.9 ± 3.8	0.28	3.5 ± 0.8	4.9 ± 5	0.12	1.1 ± 3.7	2 ± 4.7	0.6	2.9 ± 4.1 2.2 ± 3.4	0.61
Urgent triage code	14 (38%)	30 (49%)	0.44	-	12 (20%)	N/A	3 (8%)	3 (5%)	0.18	31 (84%) 11 (18%)	0.001
HFNC	7 (26%)	21 (37%)	0.18	-	5 (10%)	N/A	1 (4%)	-	N/A	4 (12%) 1 (9%)	0.76
CPAP	5 (18%)	2 (3%)	0.98	-	0	N/A	0	-	N/A	0 0	N/A
Bronchiolitis	18 (67%)	37 (65%)	0.49	-	6 (12%)	N/A	2 (9%)	-	N/A	6 (19%) 2(18%)	0.94

Abbreviation list: HFNC = high flow nasal cannula, CPAP = continuous positive airway pressure, RSV = respiratory syncytial virus, SARS-CoV-2 = severe acute respiratory syndrome coronavirus 2.

4. Discussion

Autumn and winter seasons 2022/2023 were the first seasons after the beginning of the COVID-19 pandemic during which an Influenza virus epidemic along with RSV and SARS-CoV-2 was observed in Europe [10]. This study highlighted the impact of this epidemic on urgent hospitalizations in children as related to other common viral agents.

We observed a significant growth in hospital admissions (+25%) and a higher rate of urgent hospitalizations (+51%) during autumn–winter 2022/23 compared to the previous year, and one of the determining factors appeared to be the resurgence of Influenza. The Influenza virus was responsible for many urgent hospitalizations with similar rates due to other viral agents between 2021 and 2022. Moreover, we detected a higher prevalence of hospitalizations and a more substantial impact of RSV bronchiolitis than SARS-CoV2 during autumn–winter 2022–2023. More specifically, our data aligned with other recent studies, as RSV was responsible for a greater use of respiratory support (HNFC and CPAP) in our setting compared to other viruses [11]. Whether this could suggest a more confident use of HFNC and CPAP by pediatricians outside an intensive care setting rather than an increase in severe forms of bronchiolitis is still a matter of debate. Our data support a change in the management of RSV-bronchiolitis that seems to have gained a relatively greater gravity if compared with the others respiratory infections, emerging as the primary cause of urgent hospitalization and the need for respiratory support. As demonstrated in previous studies, respiratory supports and HFNC increased significantly compared to the pre-pandemic era, but the rate of intubation and the length of stay did not change, suggesting a more aggressive treatment attitude rather than a more severe disease [12].

On the other hand, no subjects with SARS-CoV-2 positivity required respiratory aid during the last season and only one patient required it during autumn–winter 2021–2022. Our findings also aligned with previous data, showing that SARS-CoV-2 infection has a respiratory involvement significantly milder in children than in adults [13,14]. Pediatric data showed that SARS-CoV-2 only rarely causes symptoms suggestive of bronchiolitis [15]. Our data help to offer further insights into the natural history of SARS-CoV-2 in children compared to other viral agents such as Influenza and RSV. In addition, this work confirmed that respiratory support is only rarely needed in children admitted with SARS-CoV-2 infection, as already reported in previous pediatric works [15] and in extraordinary contrast with adults [16].

Finally, our data, in line with previous studies, revealed a consistent prevalence of Rhinovirus in both seasons, constituting approximately half of all viral detections. While Rhinovirus frequently co-occurs with other respiratory viral infections, our findings indicate that, despite an overall rise in its prevalence during the 2022/2023 season, the number of hospitalizations solely attributable to this virus has decreased in comparison with the previous autumn–winter season. The underlying reasons for these findings remain unclear and may be attributed to a combination of virological, environmental and behavioral factors, including the stability of these non-enveloped viruses on surfaces and their prominent transmission routes [17].

The resurgence of common viral infections was somewhat predictable, especially after the relaxation of social distancing measures, mainly school closures and facial mask wearing. On the other hand, the impact on urgent hospitalizations during the resurgence of common epidemics such as Influenza virus and RSV was less predictable. Moreover, until autumn 2022 no comparisons were possible between the Influenza virus and SARS-CoV-2.

This study has several limitations. Due to the study design, some cases may have been missed or mislabeled. Moreover, our data referred to a PED of a single Institution, so the generalizability is limited. According to the policy of our Institute, we did not perform the nasopharyngeal swab for multi-viral tests on all the patients who arrived at the PED during the study periods, but only on patients who needed urgent hospitalization. Therefore, we cannot provide the distribution of viral infections among the general population accessing the PED during the study periods. On the other hand, before the COVID-19 pandemic, the execution of viral tests in our setting was extremely rare. Therefore, we could not provide a measure of the impact of the Influenza virus and RSV on urgent hospitalizations in our setting at that time. We reported data related to the first epidemic of the Influenza virus after the pandemic, and the mean age of enrolled children was very young, so we can presume that this was the first infection with this virus for most of the children in our population, and this may have influenced the study results. No data on flu vaccination in our population were available; considering the mean age of <5 years of the children in our study, it is plausible to say that the majority of our patients were not vaccinated for SARS-CoV-2, given that vaccination was not available for this age group. Furthermore, we did not analyze the subgroup of patients with comorbidities, as these did not appear to influence the final outcome or the need for respiratory support, which was necessary in previously healthy infants with bronchiolitis.

Finally, this study was performed during the spread of the SARS-CoV-2 Delta (winter season 2021) and Omicron (winter season 2022) genetic variants throughout Italy. The cases positive for SARS-CoV-2 in this study were sustained by those variants, but no genetic data were collected.

In conclusion, this study showed that after three years of the COVID-19 pandemic, Influenza virus and RSV infections had a more significant impact on urgent hospitalizations than SARS-CoV-2. According to other Italian centers, our data may provide support for national vaccination strategies [18].

Since the beginning of the COVID-19 pandemic, remarkable scientific and public health attention has been focused on SARS-CoV-2, sometimes at the expense of measures to counteract other common viral infections in children.

Future studies are needed to maintain surveillance in the forthcoming winter season.

Author Contributions: A.A., E.B., G.C. and M.C. designed the study. F.P., B.L., C.C. and M.C. collected and analyzed the data. All the authors contributed equally to the interpretation of the results and the manuscript writing. All authors have read and agreed to the published version of the manuscript.

Funding: This research received no external funding.

Institutional Review Board Statement: The studies involving human participants were reviewed and approved by the Institutional Review Board of the Institute for Maternal and Child Health IRCCS Burlo Garofolo, Trieste, Italy.

Informed Consent Statement: Written informed consent from the participants' legal guardians/next of kin was not required to participate in this study under the national legislation and the institutional requirements.

Data Availability Statement: Data are contained within the article.

Acknowledgments: This work was supported by the Ministry of Health, Rome, Italy, in collaboration with the Institute for Maternal and Child Health IRCCS Burlo Garofolo, Trieste, Italy. The authors thank Martina Bradaschia for the English revision of the manuscript.

Conflicts of Interest: The authors declare no conflict of interest.

References

1. Amaddeo, A.; Cason, C.; Cozzi, G.; Ronfani, L.; Comar, M. Social distancing measures for COVID-19 are changing winter season. *Arch. Dis. Child.* **2021**, *106*, e47. [CrossRef] [PubMed]
2. Risso, F.M.; Cozzi, G.; Volonnino, M.; Cossovel, F.; Ullmann, N.; degli Atti, M.L.C.; Amaddeo, A.; Ghirardo, S.; Cutrera, R.; Raponi, M. Social distancing during the COVID-19 pandemic resulted in a marked decrease in hospitalisations for bronchiolitis. *Acta Paediatr.* **2022**, *111*, 163–164. [CrossRef]
3. Kuitunen, I. Influenza season 2020–2021 did not begin in Finland despite the looser social restrictions during the second wave of COVID-19: A nationwide register study. *J. Med. Virol.* **2021**, *93*, 5626–5629. [CrossRef] [PubMed]
4. Cozzi, G.; Blasutig, F.; De Nardi, L.; Giangreco, M.; Barbi, E.; Amaddeo, A. The first winter of social distancing improved most of the health indexes in a paediatric emergency department. *Acta Paediatr.* **2022**, *111*, 1027–1033. [CrossRef] [PubMed]
5. Sullivan, S.G.; Carlson, S.; Cheng, A.C.; Chilver, M.B.; Dwyer, D.E.; Irwin, M.; Kok, J.; Macartney, K.; MacLachlan, J.; Minney-Smith, C.; et al. Where has all the influenza gone? The impact of COVID-19 on the circulation of influenza and other respiratory viruses, Australia, March to September 2020. *Eurosurveillance* **2020**, *25*, 2001847. [CrossRef] [PubMed]
6. Delestrain, C.; Danis, K.; Hau, I.; Behillil, S.; Billard, M.N.; Krajten, L.; Cohen, R.; Bont, L.; Epaud, R. Impact of COVID-19 social distancing on viral infection in France: A delayed outbreak of RSV. *Pediatr. Pulmonol.* **2021**, *56*, 3669–3673. [CrossRef] [PubMed]
7. Loconsole, D.; Centrone, F.; Rizzo, C.; Caselli, D.; Orlandi, A.; Cardinale, F.; Serio, C.; Giordano, P.; Lassandro, G.; Milella, L.; et al. Out-of-season Epidemic of Respiratory Syncytial Virus during the COVID-19 Pandemic: The High Burden of Child Hospitalization in an Academic Hospital in Southern Italy in 2021. *Children* **2022**, *9*, 848. [CrossRef] [PubMed]
8. Bardsley, M.; Morbey, R.A.; Hughes, H.E.; Beck, C.R.; Watson, C.H.; Zhao, H.; Ellis, J.; Smith, G.E.; Elliot, A.J. Epidemiology of respiratory syncytial virus in children younger than 5 years in England during the COVID-19 pandemic, measured by laboratory, clinical, and syndromic surveillance: A retrospective observational study. *Lancet Infect. Dis.* **2023**, *23*, 56–66. [CrossRef] [PubMed]
9. Mrcela, D.; Markic, J.; Zhao, C.; Viskovic, D.V.; Milic, P.; Copac, R.; Li, Y. Changes following the Onset of the COVID-19 Pandemic in the Burden of Hospitalization for Respiratory Syncytial Virus Acute Lower Respiratory Infection in Children under Two Years: A Retrospective Study from Croatia. *Viruses* **2022**, *14*, 2746. [CrossRef] [PubMed]
10. Available online: https://www.cdc.gov/surveillance/resp-net/dashboard.html (accessed on 1 September 2023).
11. Brisca, G.; Mariani, M.; Buratti, S.; Ferretti, M.; Pirlo, D.; Buffoni, I.; Mallamaci, M.; Salvati, P.; Tagliarini, G.; Piccotti, E.; et al. How has the SARS-CoV-2 pandemic changed the epidemiology and management of acute bronchiolitis? *Pediatr. Pulmonol.* **2023**, *58*, 1169–1177. [CrossRef] [PubMed]
12. Ghirardo, S.; Cozzi, G.; Tonin, G.; Risso, F.M.; Dotta, L.; Zago, A.; Lupia, D.; Cogo, P.; Ullmann, N.; Coretti, A.; et al. Increased use of high-flow nasal cannulas after the pandemic in bronchiolitis: A more severe disease or a changed physician's attitude? *Eur. J. Pediatr.* **2022**, *181*, 3931–3936. [CrossRef] [PubMed]
13. Woolf, S.H.; Chapman, D.A.; Lee, J.H. COVID-19 as the Leading Cause of Death in the United States. *JAMA* **2021**, *325*, 123–124. [CrossRef] [PubMed]
14. Servidio, A.G.; Visentin, G.; Conti, R.; Cozzi, G.; Travan, L.; Bua, J.; Barbi, E.; Amaddeo, A. Mild COVID-19 in hospitalised infants younger than 90 days. *Acta Paediatr.* **2023**, *112*, 483–485. [CrossRef] [PubMed]
15. Cozzi, G.; Cozzi, G.; Wiel, L.C.; Wiel, L.C.; Amaddeo, A.; Amaddeo, A.; Gatto, A.; Gatto, A.; Giangreco, M.; Giangreco, M.; et al. Prevalence of SARS-CoV-2 positivity in infants with bronchiolitis: A multicentre international study. *Arch. Dis. Child.* **2022**, *107*, 840–844. [CrossRef] [PubMed]
16. Hasnat, F.; Noman, F.; Moben, A.L.; Sarker, A.S.; Morshed, J.; Mutanabbi, M. Difference in Clinical Patterns between COVID-19 Affected Children and Adults. *Mymensingh Med. J.* **2021**, *30*, 1093–1099. [PubMed]
17. Rankin, D.A.; Spieker, A.J.; Perez, A.; Stahl, A.L.; Rahman, H.K.; Stewart, L.S.; Schuster, J.E.; Lively, J.Y.; Haddadin, Z.; Probst, V.; et al. Circulation of Rhinoviruses and/or Enteroviruses in Pediatric Patients With Acute Respiratory Illness Before and During the COVID-19 Pandemic in the US. *JAMA Netw. Open* **2023**, *6*, e2254909. [CrossRef] [PubMed]
18. Ciofi Degli Atti, M.; Rizzo, C.; D'Amore, C.; Ravà, L.; Reale, A.; Barbieri, M.A.; Bernaschi, P.; Russo, C.; Villani, A.; Perno, C.F.; et al. Acute respiratory infection emergency access in a tertiary care children hospital in Italy, prior and after the SARS-CoV-2 emergence. *Influenza Other Respir. Viruses* **2023**, *17*, e13102. [CrossRef] [PubMed]

viruses

MDPI

Article

Respiratory Syncytial Virus-Associated Deaths among Children under Five before and during the COVID-19 Pandemic in Bangladesh

Md Zakiul Hassan [1,*], Md. Ariful Islam [1], Saleh Haider [1], Tahmina Shirin [2] and Fahmida Chowdhury [1]

1 International Centre for Diarrhoeal Disease Bangladesh (icddr,b), Dhaka 1213, Bangladesh; arif@icddrb.org (M.A.I.); abu.haider@icddrb.org (S.H.); fahmida_chow@icddrb.org (F.C.)
2 Institute of Epidemiology, Disease Control and Research (IEDCR), Dhaka 1212, Bangladesh; tahmina.shirin14@gmail.com
* Correspondence: zhassan@icddrb.org

Abstract: Respiratory syncytial virus (RSV) is a leading cause of acute lower respiratory infections in young children worldwide. RSV-associated deaths in children are underreported in Bangladesh. We analyzed hospital-based surveillance data on severe acute respiratory infections (SARIs) in under-five children before (August 2009–February 2020) and during the COVID-19 pandemic (March 2020–March 2022). Using the World Health Organization definition, we identified SARI cases in 14 tertiary-level hospitals. Nasopharyngeal and oropharyngeal swabs were collected for real-time reverse-transcriptase–polymerase chain reaction (rRT-PCR) testing of six respiratory viruses, including RSV. SARI deaths during the pandemic (2.6%, 66) were higher than pre-pandemic (1.8%, 159; $p < 0.001$). Nearly half of pandemic deaths (47%) had underlying respiratory viruses, similar to the pre-pandemic rate (45%). RSV detection in deaths was consistent pre-pandemic (13%, 20/159) and during the pandemic (12%, 8/66). Children aged < 6 months constituted 57% (16) of RSV-related deaths. Evaluating interventions like maternal vaccination and infant monoclonal antibody prophylaxis is crucial to address RSV, a major contributor to under-five SARI deaths.

Keywords: respiratory syncytial virus; RSV; COVID-19 pandemic; children; SARI death; Bangladesh

Citation: Hassan, M.Z.; Islam, M.A.; Haider, S.; Shirin, T.; Chowdhury, F. Respiratory Syncytial Virus-Associated Deaths among Children under Five before and during the COVID-19 Pandemic in Bangladesh. *Viruses* **2024**, *16*, 111. https://doi.org/10.3390/v16010111

Academic Editors: Emanuele Castagno and Irene Raffaldi

Received: 11 November 2023
Revised: 12 December 2023
Accepted: 15 December 2023
Published: 12 January 2024

1. Introduction

Respiratory syncytial virus (RSV) is a major cause of respiratory tract infections worldwide in infants and young children [1]. It is highly contagious and transmitted through respiratory droplets and touch. RSV-infected individuals often exhibit symptoms 4–6 days after infection. The symptoms generally include sneezing, coughing, wheezing, runny nose, increase in temperature, and decrease in appetite. Typically, these symptoms do not appear together; rather, they appear gradually. Infected newborns and infants may only symptomize diminished activity or lethargy, irritability, and breathing problems. Nearly every child faces RSV exposure before their second birthday [2]. Although most of the infected children are cured without any complications, some develop severe infections. The most common complications include bronchiolitis or inflammation of the small airways of the lungs, pneumonia, and eye and ear infections [2]

RSV poses a significant global health burden [3], predominantly in low- and middle-income countries (LMICs) [4]. Globally, 1 in every 50 deaths among under-five children is attributable to RSV, and in 2019 alone, there were around 3.6 million hospital admissions and 26,300 in-hospital deaths globally in children under five years due to RSV [4]. LMICs bear the brunt of the RSV burden, accounting for over 95% of RSV-acute lower respiratory infections and over 97% of related mortality in all age groups globally [4].

Bangladesh, an LMIC, faces a significant RSV burden, with approximately 90% of excess mortality during the RSV season attributed to RSV infections [5]. Hospitalization rates due to RSV are considerable, with RSV having the highest burden among both hospitalized (34%) and non-hospitalized (39%) cases in children under the age of five [6]. The economic impact of RSV is also substantial, with families spending a median direct cost of USD 62 and an indirect cost of USD 19 for hospitalization in 2010 [7]. The estimated median direct cost of RSV-associated hospitalization in children under five years was USD 10 million, with an indirect cost of USD 3.0 million in 2010, placing significant financial strain on affected families and the healthcare system [7].

The COVID-19 pandemic brought about significant changes in the epidemiology of RSV worldwide. Studies in developed countries reported delayed RSV peaks during the pandemic following an absence during its typical season [8]. Reduced transmission of RSV was observed due to non-pharmaceutical interventions, like lockdowns and school closings [9]. Furthermore, changes in the age distribution of RSV infections have been noted, with preschool-aged children being more affected than school-going children and newborns [10]. However, the specific impact of the pandemic on RSV epidemiology in LMICs, including Bangladesh, remains an important data gap that needs to be addressed.

Understanding the burden of RSV-associated under-five child deaths in Bangladesh, both during and before the COVID-19 pandemic, is critical for designing context-specific interventions and public health policies. Data on RSV-associated morbidity and mortality may guide effective strategies to reduce RSV burden. This study aimed to address these gaps and generate critical evidence on RSV-associated deaths among under-five children with SARI in Bangladesh, ultimately contributing to improved public health responses and better outcomes for the vulnerable under-five population.

2. Materials and Methods

2.1. Hospital-Based Surveillance Platform

We analyzed the data from the hospital-based influenza surveillance (HBIS) system to characterize SARI deaths. As a part of the National Influenza Centres (NIC), the HBIS was initiated in Bangladesh in 2007 in 12 tertiary care hospitals. The surveillance was conducted in a maximum of 14 hospitals at different geographical locations across Bangladesh over different time points (Figure 1); the number of sites ranged from 7 to 14 depending on various years. Since 2018, it has been operational in nine tertiary-care-level hospitals (seven public and two private) geographically distributed all over Bangladesh. The activities of the surveillance system are carried out jointly by the International Centre for Diarrhoeal Disease Research, Bangladesh (icddr,b) and the Institute of Epidemiology, Disease Control and Research (IEDCR) of the Government of Bangladesh (GoB), with technical support from the United States Centers for Disease Control and Prevention (US CDC). The inpatient capacity of the surveillance hospitals ranges from 500 to 1500 beds, with a 100–150% bed occupancy rate.

2.2. Study Population

For this study, we analyzed the data of participants enrolled from August 2009 to March 2022 in HBIS. Despite the ongoing pandemic in 2020 and subsequent pandemic control efforts, the surveillance remained active and continued its operations six days a week (Saturday to Thursday) during working hours (8:30 a.m. to 5:00 p.m.) by collecting data from the in-patient departments of the study hospitals. During the national holidays and the weekends (Friday), data collection was paused. In our study, the pre-pandemic period spanned from August 2009 to February 2020, and the pandemic period was considered from March 2020 to March 2022. Our study population contains all in-patient children under 5 years old admitted to the surveillance facilities during these periods.

2.3. Case Identification

The surveillance physicians and support staff screened and identified the severe acute respiratory infection (SARI) patients who met the WHO case definition of SARI, defined as an acute respiratory infection with subjective or measured fever of $\geq$38 °C and a history of cough with onset within the last 10 days from in-patient departments of medicine and pediatrics wards, coronary care units (CCUs), and specialized COVID-19 isolation wards established during the COVID-19 pandemic.

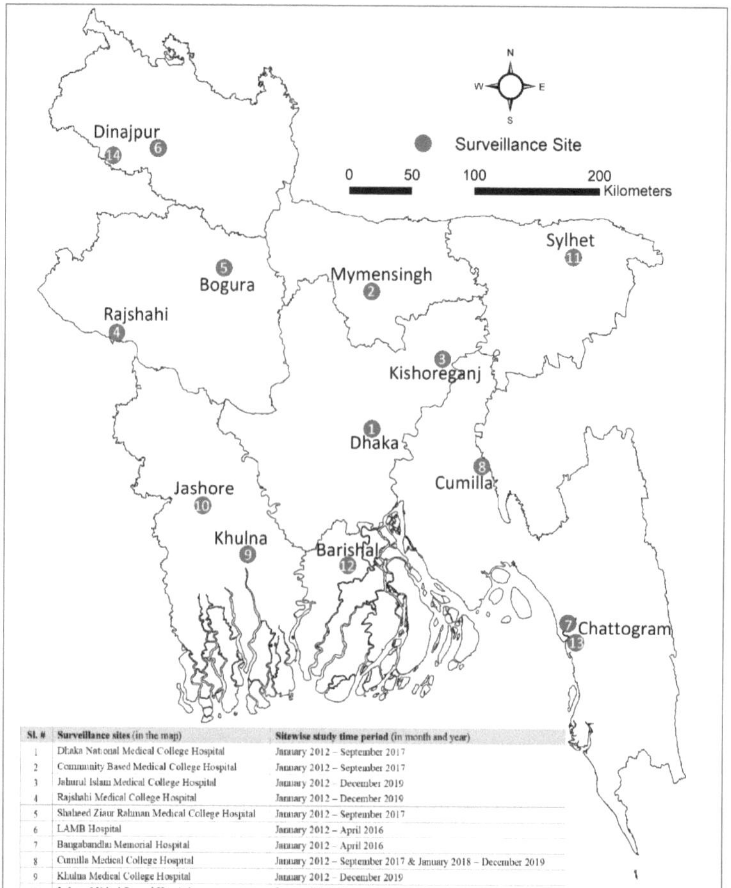

Sl. #	Surveillance sites (in the map)	Sitewise study time period (in month and year)
1	Dhaka National Medical College Hospital	January 2012 – September 2017
2	Community Based Medical College Hospital	January 2012 – September 2017
3	Jahurul Islam Medical College Hospital	January 2012 – December 2019
4	Rajshahi Medical College Hospital	January 2012 – December 2019
5	Shaheed Ziaur Rahman Medical College Hospital	January 2012 – September 2017
6	LAMB Hospital	January 2012 – April 2016
7	Bangabandhu Memorial Hospital	January 2012 – April 2016
8	Cumilla Medical College Hospital	January 2012 – September 2017 & January 2018 – December 2019
9	Khulna Medical College Hospital	January 2012 – December 2019
10	Jashore 250 bed General Hospital	January 2012 – September 2017 & December 2018 – December 2019
11	Jalalabad Ragib-Rabeya Medical College Hospital	January 2012 – December 2019
12	Sher-e-Bangla Medical College Hospital	January 2012 – December 2019
13	Chattogram Medical College Hospital	April 2014 – December 2019
14	M Abdur Rahim Medical College Hospital	April 2014 – December 2019

Figure 1. Location of the 14 surveillance hospitals of HBIS in Bangladesh.

2.4. *Data Collection*

After identification, written informed consent was obtained from the parents or caregivers of the under-five children with SARI. The surveillance physicians then enrolled and performed a physical examination of all the under-five children with SARI. This was followed by the collection of data using a standardized surveillance form on a handheld computer. Real-time data transfers to our central server allowed for the development of algorithms that reported on primary missing variables and/or values in variables related to data quality. The form included demographic, clinical, history of comorbid conditions (e.g., diabetes, hypertension, cancer, asthma, chronic obstructive pulmonary disease, heart diseases), and available diagnostic findings of the patients. At the time of discharge, the outcome status (referral to another facility, partial recovery, full recovery, and in-hospital death) of the participants was recorded. The data were checked for a second time by the data management team and matched with the staff to ensure the rectification of the inaccuracies and to establish a robust system.

2.5. Sample Collection and Transportation

Maintaining all aseptic precautions, nasopharyngeal (NP) and oropharyngeal (OP) swabs were collected from the SARI patients after written informed consent from their parents or caregivers to participate in our study. Swabs were put into individual cryovials containing VTM and kept in a cool box for up to 30 min with a temperature between 2 °C and 8 °C. Both the NP swab and OP swab samples were labeled, packaged, stored in a nitrogen dry shipper (−150 °C) on site, and then transported to the virology laboratory of icddr,b, Dhaka, every week.

2.6. Laboratory Analysis

We tested all the in-hospital death cases for common respiratory viruses: RSV, adenoviruses, influenza, human metapneumovirus (HMPV), human parainfluenza viruses (HPIV), and SARS-CoV-2 (from March 2020); these were checked using real-time reverse-transcriptase–polymerase chain reaction (rRT-PCR). At icddr,b virology laboratory, InviMag Virus DNA/RNA Mini Kit (Invitek, STRATEC Molecular, Berlin, Germany) was used on the Kingfisher Flex 96 (Thermo Fisher Scientific, Waltham, MA, USA), an automated nucleic acid extraction tool to extract the viral nucleic acid from pooled NP and OP swab samples [11]. US-CDC provided us with the primers and probes. The primers and probes of the rRT-PCR assay were designed to detect the conserved regions of matrix genes. These are detected from the GenBank alignment sequences. The primers and probes used for detection of the the viruses of interest via rRT-PCR assay are given in the table below (Table 1) [12,13]. Specimen total nucleic acid (TNA) extract (5 µL), forward primer (0.5 µM), reverse primer (0.25 µM), and probe (0.05 µM) were used to prepare the reaction mixture. It was then amplified using an iCycler iQ^{TM} Real-Time Detection System (Bio-Rad, Hercules, CA, USA). Three cycling conditions were used: 1 10 min cycle at 48 °C, 1 5 min cycle at 95 °C, and then 45 cycles of 15 s at 95 °C followed by 1 min at 55 °C [14].

Table 1. Primers and probes used in this study for real-time reverse-transcriptase–polymerase chain reaction (rRT-PCR) assays.

Assay	Primer/Probe Sequence (5′–3′)
RSV	F, GGC AAA TAT GGA AAC ATA CGT GAA R, TCT TTT TCT AGG ACA TTG TAY TGA ACA G P, CTG TGT ATG TGG AGC CTT CGT GAA GCT
Influenza A	F, GAC CRA TCC TGT CAC CTC TGA C R, AGG GCA TTY TGG ACA AAK CGT CTA P, TGC AGT CCT CGC TCA CTG GGC ACG
Influenza B	F, TCC TCA ACT CAC TCT TCG AGC G R, CGG TGC TCT TGA CCA AAT TGG P, CCA ATT CGA GCA GCT GAA ACT GCG GTG
SARS-CoV-2	F, GACCCCAAAATCAGCGAAAT R, TCTGGTTACTGCCAGTTGAATCTG P, ACCCCGCATTACGTTTGGTGGACC
HMPV	F, CAA GTG TGA CAT TGC TGA YCT RAA 2 R, ACT GCC GCA CAA CAT TTA GRA A P, TGG CYG TYA GCT TCA GTC AAT TCA ACA GA
Adenovirus	F, GCC CCA GTG GTC TTA CAT GCA CAT C 2 R, GCC ACG GTG GGG TTT CTA AAC TT P, TGC ACC AGA CCC GGG CTC AGG TAC TCC GA
HPIV	F, AGT TGT CAA TGT CTT AAT TCG TAT CAA T 2 R, TCG GCA CCT AAG TAA TTT TGA GTT P, ATA GGC CAA AGA "T"TG TTG TCG AGA CTA TTC CAA

F, forward primer; R, reverse primer; P, probe; (P) = dP-CE (pyrimidine derivative); (A) = LNA-dA, (T) = LNA-dT (Locked Nucleic Acid (LNA) primers).

2.7. Data Analysis

We calculated descriptive statistics for all variables. Continuous variables were summarized using median and interquartile range (IQR) based on the distribution of the variables. We provided frequencies and proportions for categorical variables. We also used Chi-square and Fisher's exact tests to compare the contribution of RSV in SARI mortality among under-5 children before and during the pandemic, where we considered a p-value < 0.05 statistically significant. We conducted the statistical analyses using Stata version 15, College Station, TX 77845, USA.

3. Results

3.1. Demographics of Study Participants

We enrolled 8923 under-5-year-old children with SARI during the pre-COVID-19 pandemic phase (August 2009–February 2020). The median age was 6 months (IQR: 2.5–12), and 67% were male (5956). During the pandemic period (March 2020–March 2022), 2570 children < 5 years were enrolled. The median age was found to be 6 months (IQR: 3–14): 65% males (1680). Almost 90% of the patients were younger than two years throughout the whole study period (Table 2).

Table 2. Demographic, clinical and epidemiological characteristics of under-five children with severe acute respiratory infections (SARIs) before the COVID-19 pandemic (August 2009–February 2020) and during the COVID-19 pandemic (March 2020–March 2022) in Bangladesh.

Characteristics	SARI Patients Enrolled			p-Value
	Total SARI Patients N = 11,493 *n* (%)	**Before Pandemic N = 8923 *n* (%)**	**During Pandemic N = 2570 *n* (%)**	
Demographic characteristics				
Age				
<2 Year	10,238 (89)	7980 (89.4)	2258 (88)	0.059
2–5 Years	1255 (11)	943 (10.6)	312 (12)	<0.001
Median age (IQR), years	0.5 (0.2–1)	0.5 (0.2–1)	0.6 (0.2–1.2)	<0.001
Sex (Male)	7637 (66.4)	5957 (66.8)	1680 (65.4)	0.188
Clinical Characteristics				
Runny nose	6925 (60.3)	5076 (57)	1849 (72)	<0.001
Difficulty of breathing	9959 (86.7)	7769 (87)	2190 (85)	0.014
Sore throat	51 (0.4)	51 (0.6)	0 (0)	-
Chest indrawing	8760 (76.2)	7126 (80)	1634 (63.6)	<0.001
Unable to drink	2835 (24.7)	2080 (23.3)	755 (29.4)	<0.001
Vomiting	1571 (13.7)	1446 (16.2)	125 (5)	<0.001
Lethargy	668 (5.8)	628 (7)	40 (1.6)	<0.001
Diarrhea	227 (2)	181 (2)	46 (1.8)	0.443
Duration of symptoms prior to admission in days; Median (IQR)	2 (1–3)	2 (1–3)	2 (1–3)	0.036
Length of hospital stay in days; Median (IQR)	4 (3–6)	4 (3–6)	4 (2–6)	<0.001
Co-morbid condition				
≥1 co-morbid condition (Self-reported)	11,406 (99.2)	8855 (99.2)	2551 (99.3)	0.907
Treatment received				
Antibiotic	7804 (91)	5398 (90)	2406 (93.6)	<0.001
Oseltamivir	0 (0)	0 (0)	0 (0)	-
Oxygen	3458 (41)	2232 (37.7)	1226 (47.7)	<0.001
Mechanical ventilation	4 (0.05)	4 (0.07)	0 (0)	-

Table 2. *Cont.*

Characteristics	SARI Patients Enrolled			
	Total SARI Patients N = 11,493 *n* **(%)**	**Before Pandemic N = 8923** *n* **(%)**	**During Pandemic N = 2570** *n* **(%)**	***p*-Value**
ICU support (after admission in general ward)	9 (0.1)	8 (0.1)	1 (0.4)	-
Laboratory Results				
RSV †	658 (29)	638 (29)	20 (17)	0.242
Influenza virus	898 (7.8)	698 (7.8)	200 (7.8)	1
SARS-CoV-2 ᵡ	6 (0.23)	0 (0)	6 (0.23)	-
HMPV †	168 (7.3)	164 (7.5)	4 (3.4)	0.757
Adenovirus †	155 (6.7)	141 (6.4)	14 (12)	0.43
HPIV †	159 (6.9)	147 (6.7)	11 (9.4)	0.733
Co-detection with ≥2 respiratory viruses	150 (1.3)	136 (1.5)	14 (0.5)	0.762
Clinical outcome; Death	225 (2)	159 (1.8)	66 (2.6)	<0.001

† A total of 2307 samples were tested for RSV, HMPV, Adenovirus and HPIV. ᵡ A total of 2563 samples collected during March 2020–March 2022 were tested for SARS-CoV-2.

3.2. Clinical Features of Study Participants

During both of the periods, the most common respiratory symptoms the patients reported were breathing difficulty (87%) and chest indrawing (76%). Other common clinical symptoms included runny nose and inability to drink. The parents or the caregivers reported to the hospital within an average of 2 days (IQR: 1–3 days) of the onset of symptoms. Almost all patients were discharged within 4 days (IQR: 3–6 days) of admission unless they were referred to a different facility or died.

3.3. Clinical Care of Study Participants

Over 90% of the children received antibiotics. None received any antiviral drugs. The necessity of supplementary oxygen increased more during the pandemic than in the pre-pandemic period. A total of nine patients required ICU support, and only one of them was from the pandemic period (Table 2).

3.4. Contribution of RSV in SARI Mortality Pre- and during Pandemic

We found that compared to the pre-pandemic period, the proportion of SARI deaths at the time of the pandemic was higher [((1.8%, 159) vs. 2.6%, 66); $p < 0.001$]. During the pre-pandemic period, 45% (71/159) of the death cases revealed respiratory viruses, whereas it was 47% (31/66) during the pandemic (Figure 2). Of 159 pre-pandemic deaths, RSV was predominantly detected (13%, 20). The other findings included adenovirus (8%, 12), HPIV (9%, 14), HMPV (6%, 10), and influenza (4%, 6). Further, nine (6%) death cases detected viral co-detections, including three (2%) co-detections with RSV. During the pandemic period, RSV (12%, 8) as well as adenovirus (12%, 8) comprised the largest proportions of the 66 pandemic deaths. The other detected viruses were SARS-CoV-2 (6%, 4), HPIV (6%, 4), influenza (3%, 2), and HMPV (1%, 2). We also detected co-detection with RSV and adenovirus (3%, 2) and with HPIV and adenovirus (3%, 2) (Figure 3).

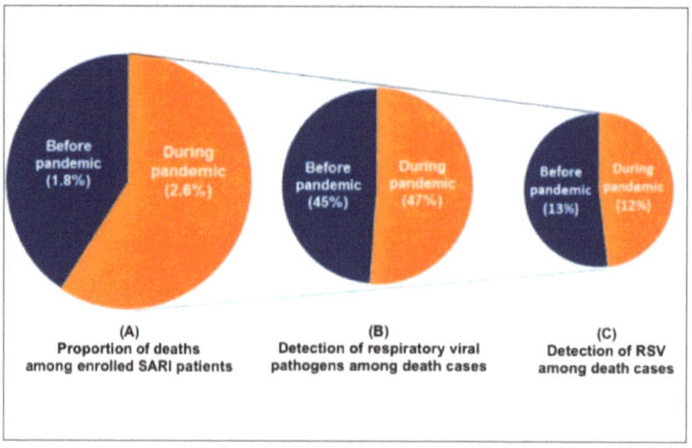

Figure 2. Proportion of deaths among the SARI patients aged < 5 years before and during the pandemic.

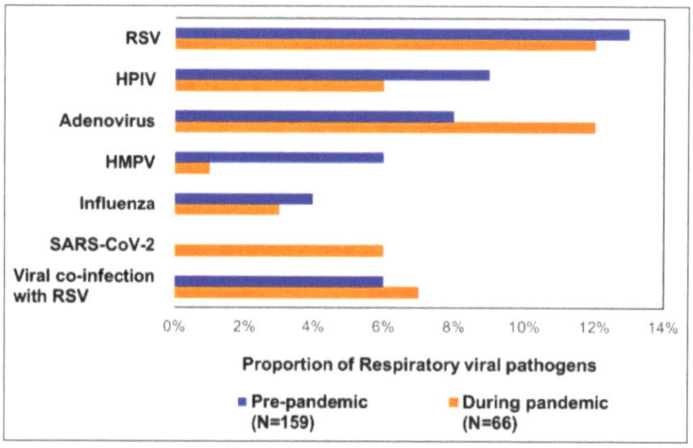

Figure 3. Respiratory viral pathogens detected among SARI death cases (aged $\leq$ 5 years) during the pre-pandemic and pandemic periods.

3.5. Characteristics of RSV-Associated SARI Deaths among under-Five Children before and during the COVID-19 Pandemic

Before the onset of the COVID-19 pandemic, there were 20 recorded deaths in children under the age of 5 years attributed to severe acute respiratory syndrome associated with RSV. The median age of these cases was 3.5 months (IQR: 2.3–6 months), and 65% of the cases were male (13/20). Of these cases, 85% (17/20) were children aged < 1 year and 65% (13/20) were aged < 6 months. Co-morbid conditions were present in 10% of the cases (2/20).

During the COVID-19 pandemic, there were eight deaths in under-5-year-old children associated with RSV-SARI. The median age for these cases was 7.5 months (IQR: 2.5–13.5 months), and 25% were male (2/8). Among these cases, 75% (6/8) were children under 1 year of age, 38% (3/8) were under 6 months of age, and 13% (1/8) had at least one co-morbid condition.

RSV was solely detected in 57% (16) of the <6 months old, in 25% (7) of the 6–12 months old, in 11% (3) 1–2 years old, and 7% (2) of the 3–5 year olds among all the death cases. All cases exhibited a history of breathing difficulty.

4. Discussion

Our study showed that in both the pre-pandemic and pandemic periods, RSV was the major contributor to deaths among young children with SARI. Notably, we observed that nearly half of all SARI-related deaths in under-5-year-old children were associated with various respiratory viral pathogens, with RSV consistently responsible for a substantial proportion of these cases, exhibiting minimal variation between the two periods (pre-pandemic: 13%; during pandemic: 12%). These findings underscore the urgency of implementing measures to prevent these vaccine-preventable deaths.

The burden and impact of RSV infections among under-five children have not been well studied in Bangladesh. Our study, as far as we are aware, represents the first of its kind in Bangladesh, concentrating exclusively on mortality among children under the age of five linked to RSV. One of the main causative agents behind Bronchiolitis is RSV. According to the WHO estimates, RSV is responsible for about 60% of pediatric acute respiratory infections worldwide. Moreover, during the height of the viral season each year, RSV causes about 80% of lower respiratory tract infections (LRTIs) in infants under the age of one [15]. Pneumonia, another severe LRTI complication of RSV, may occur with or without bacterial co-detection. Around 40% of those who were admitted into the pediatric intensive care unit (PICU) due to severe RSV bronchiolitis were co-infected with respiratory bacteria and had a higher risk for bacterial pneumonia [16]. A prior study conducted in Bangladesh revealed a surge in unclassified pneumonia-related deaths (64%) in children < 2 years during the peak seasons of bronchiolitis [17].

Estimating RSV-related mortality in Bangladesh has proven challenging, primarily due to the absence of comprehensive, long-term systematic surveillance [18]. Our ongoing surveillance platform, the HBIS, has been consistently monitoring data, both before and after the onset of the COVID-19 pandemic. This platform enables us to systematically analyze deaths attributed to RSV and all respiratory viruses in a comprehensive manner. The findings from our study underscore the necessity for continued surveillance and further research to investigate the underlying factors associated with RSV-related mortality and to identify effective strategies for preventing childhood RSV infections and deaths in Bangladesh.

This study reveals a slight decrease in the total number of RSV-associated deaths among children under five years old with SARI during the COVID-19 pandemic. Previously, RSV-related deaths accounted for 20 out of 159 (13%) of all deaths, but during the pandemic, this figure slightly decreased to 8 out of 66 (12%). It is worth noting that the total number of SARI patients in children under five also decreased during this period. This trend of reduced RSV-associated morbidity and mortality during the pandemic has been observed in several other countries, including China [19], Austria [20], France [21], Brazil [22], and globally [23–25].

Apart from the concept of viral interference, the non-pharmacological interventions (NPIs) enacted during the COVID-19 pandemic, such as the use of face masks, frequent handwashing, social distancing, and lockdowns, may have played a role in the lower circulation of RSV. Washing hands frequently, putting on masks, and practicing respiratory hygiene habits helped to stop the spread of RSV along with COVID-19. These practices limited the circulation of virus-containing respiratory droplets. Curbing large gatherings and physical distancing also limited viral transmission. In addition to helping to avoid COVID-19, this action reduced the community spread of RSV [26].

When compared to the pre-pandemic period, our research indicates a shift in the median age of RSV-related deaths among children under five, increasing from 3.5 months (IQR 2.3–6) to 7.5 months (IQR 2.5–13.5) during the pandemic. The proportion of deaths due to RSV among children aged under six months decreased from 65% before the pandemic to 38% during the pandemic. This suggests that a higher percentage of older children, primarily over 6 months old, were succumbing to RSV infections during the pandemic compared to the pre-pandemic period. A study in France also reported a similar increase in the median age of RSV admissions among children [27]. An Australian study found that the median age of RSV-associated hospitalization in children significantly rose from 12.5 months in 2019 to 18.4 months after the COVID-19 pandemic NPIs were relaxed [28]. Various studies worldwide have reported that the decline in or near absence of RSV cases was followed by a delayed seasonal resurgence, accompanied by an increase in the median age of infection and death. We attribute this phenomenon to a significant cohort of older children who remained immunologically vulnerable due to the NPIs during the COVID-19 pandemic, and later, when these measures were relaxed, they came into contact with the virus [19,29–31].

To combat emerging RSV epidemics, it is necessary to promote personal hygiene practices and social distancing for sick individuals. It is neither practical nor feasible to maintain year-round generalized social distancing and lockdowns. But season, context, population-specific (e.g., kindergarten, elementary schools, etc.) mask mandates, temporary restrictions on large gatherings, and distancing should be taken into consideration as proven methods to curb the burden of RSV as well as all respiratory viral infections during future potential outbreaks.

Among other preventive measures, infant monoclonal antibody prophylaxis and RSV vaccination for mothers are noteworthy. Immunizing the pregnant mother during the second or third trimester of the pregnancy will boost the serum-neutralizing antibody response, the serum of which will eventually transfer to the fetus from the mother via the placenta during the prenatal period and via breastmilk during the postnatal period [32,33]. These maternal antibodies effectively produce immunity against RSV in RSV-naïve newborns and infants. The monoclonal antibodies provide pre-RSV exposure or prophylactic and passive immunization to infants, especially preterm or full-term at-risk babies from severe lower respiratory tract infection at the early stage of their lives [34]. These two options should also be evaluated to combat these RSV-associated premature deaths.

We observed that the presence of co-morbid conditions was 10% (2/20) before the pandemic and 13% (1/8) during the pandemic. Notably, a history of breathing difficulty was a common feature among children under five upon hospitalization. Due to the limited number of deaths, conducting a risk factor analysis was not feasible. However, future work should consider such analysis using a larger cohort and develop risk detection or prediction models to predict unfavorable outcomes. In resource-constrained settings like ours, the creation and utilization of clinical prediction tools can facilitate early disease severity detection, aid in diagnosis and prognosis, and offer clinical decision support [35,36].

Our study has several limitations. Firstly, we only tested SARI death cases. To gain a better understanding of the seasonality of RSV and the impact of COVID-19 on the seasonality and circulation of RSV, we need continuous year-round geographically representative surveillance. Secondly, we only used data from our hospital-based SARI surveillance and likely missed RSV cases that are non-SARI or non-medically attended due to healthcare-seeking behavior, where only 34% of the population receives healthcare from trained medical personnel [37]. We were also unable to conduct a risk factor analysis due to the low number of deaths captured through the hospital-based SARI surveillance system. Community-based surveillance can provide a more accurate insight into the RSV circulation and burden in Bangladesh.

5. Conclusions

RSV imposes a significant burden among under-five children, particularly in LMICs. Despite being a leading cause of under-five mortality in Bangladesh, the RSV burden has not been well studied. Our study has shown that both before and during the pandemic periods, RSV was a significant factor, leading to mortality in under-five children with SARI in Bangladesh. Despite some changes in RSV circulation due to the COVID-19 pandemic, it is expected that RSV circulation will re-emerge and cause local outbreaks in the near future. An increase in the median age of RSV-infected children indicated naturally unimmunized older children became vulnerable with the relaxation of NPIs.

Hospital- and community-based systematic surveillance is important to monitor RSV circulation and characterize RSV seasonality. National health authorities should promote personal respiratory hygiene. Season-, context-, and population-specific mask mandates and temporary social distancing should be implemented to minimize the community spread of RSV during potential seasonal outbreaks. In addition to that, other evidence-based measures, such as monoclonal antibody prophylaxis for infants and RSV vaccination for mothers, should be evaluated. These preventive interventions may help us combat RSV-associated unexpected premature deaths in the future.

Author Contributions: Conceptualization, M.Z.H., M.A.I. and F.C.; methodology, M.Z.H., M.A.I., S.H., T.S. and F.C.; software, M.Z.H., M.A.I. and S.H.; validation, M.Z.H., M.A.I., S.H., T.S. and F.C.; formal analysis, M.Z.H. and M.A.I.; investigation, M.Z.H., M.A.I., S.H., T.S. and F.C.; resources, M.Z.H. and M.A.I.; data curation, M.Z.H. and M.A.I.; writing—original draft preparation, M.Z.H. and S.H.; writing—review and editing, M.A.I., S.H., T.S. and F.C.; visualization, M.Z.H. and M.A.I.; supervision, T.S. and F.C.; project administration, M.Z.H., M.A.I. and F.C.; funding acquisition, M.Z.H., M.A.I. and F.C. All authors have read and agreed to the published version of the manuscript.

Funding: This research was funded by the Centers for Disease Control and Prevention (CDC), Atlanta, Georgia, USA, grant number [U01GH002259]. The APC was funded by the US CDC.

Institutional Review Board Statement: The study was conducted in accordance with the Declaration of Helsinki and approved by the Institutional Review Board (or Ethics Committee) of the Institutional Review Board of icddr,b.

Informed Consent Statement: The legal guardians of the under-five children provided written informed consent.

Data Availability Statement: The data presented in this study are available on request from the corresponding author. The data are not publicly available to ensure the protection of privacy.

Acknowledgments: We acknowledge surveillance physicians, collaborators at the hospitals, and field staff for their continuous support. We are grateful to the study participants for their time. Hospital-based influenza surveillance was funded by the Centers for Disease Control and Prevention (CDC), Atlanta, USA, under cooperative agreement number U01GH002259. icddr,b acknowledges with gratitude the commitment of CDC to its research efforts. icddr,b is grateful to the Governments of Bangladesh and Canada for providing core/unrestricted support.

Conflicts of Interest: The authors declare no conflict of interest.

References

1. Hall, C.B. Respiratory Syncytial Virus and Parainfluenza Virus. *N. Engl. J. Med.* **2001**, *344*, 1917–1928. [CrossRef] [PubMed]
2. CDC. Respiratory Syncytial Virus Infection (RSV). 2023. Available online: https://www.cdc.gov/rsv/about/symptoms.html (accessed on 20 November 2023).
3. Hall, C.B. Nosocomial Respiratory Syncytial Virus Infections: The "Cold War" Has Not Ended. *Clin. Infect. Dis.* **2000**, *31*, 590–596. [CrossRef] [PubMed]
4. Li, Y.; Wang, X.; Blau, D.M.; Caballero, M.T.; Feikin, D.R.; Gill, C.J.; Madhi, S.A.; Omer, S.B.; Simoes, E.A.F.; Campbell, H.; et al. Global, regional, and national disease burden estimates of acute lower respiratory infections due to respiratory syncytial virus in children younger than 5 years in 2019: A systematic analysis. *Lancet* **2022**, *399*, 2047–2064. [CrossRef] [PubMed]
5. Shi, T.; McAllister, D.A.; O'Brien, K.L.; Simoes, E.A.F.; Madhi, S.A.; Gessner, B.D.; Polack, F.P.; Balsells, E.; Acacio, S.; Aguayo, C.; et al. Global, regional, and national disease burden estimates of acute lower respiratory infections due to respiratory syncytial virus in young children in 2015: A systematic review and modelling study. *Lancet* **2017**, *390*, 946–958. [CrossRef] [PubMed]
6. Nasreen, S.; Luby, S.P.; Brooks, W.A.; Homaira, N.; Al Mamun, A.; Bhuiyan, M.U.; Rahman, M.; Ahmed, D.; Abedin, J.; Rahman, M.; et al. Population-based incidence of severe acute respiratory virus infections among children aged <5 years in rural Bangladesh, June–October 2010. *PLoS ONE* **2014**, *9*, e89978. [CrossRef]
7. Bhuiyan, M.U.; Luby, S.P.; Alamgir, N.I.; Homaira, N.; Sturm-Ramirez, K.; Gurley, E.S.; Abedin, J.; Zaman, R.U.; Alamgir, A.; Rahman, M.; et al. Costs of hospitalization with respiratory syncytial virus illness among children aged <5 years and the financial impact on households in Bangladesh, 2010. *J. Glob. Health* **2017**, *7*, 010412. [CrossRef] [PubMed]
8. Agha, R.; Avner, J.R. Delayed Seasonal RSV Surge Observed during the COVID-19 Pandemic. *Pediatrics* **2021**, *148*, e2021052089. [CrossRef]
9. Chow, E.J.; Uyeki, T.M.; Chu, H.Y. The effects of the COVID-19 pandemic on community respiratory virus activity. *Nat. Rev. Microbiol.* **2023**, *21*, 195–210. [CrossRef]
10. Zhang, M.; Gao, J.; Guo, Q.; Zhang, X.; Zhang, W. Changes of respiratory syncytial virus infection in children before and after the COVID-19 pandemic in Henan, China. *J. Infect.* **2023**, *86*, 154–225. [CrossRef]
11. Healthineers, S. Molecular Diagnostics. Available online: http://www.fast-trackdiagnostics.com/human-line/products/ftd-respiratory-pathogens-33/ (accessed on 20 November 2023).

12. Lu, X.; Wang, L.; Sakthivel, S.K.; Whitaker, B.; Murray, J.; Kamili, S.; Lynch, B.; Malapati, L.; Burke, S.A.; Harcourt, J.; et al. US CDC Real-Time Reverse Transcription PCR Panel for Detection of Severe Acute Respiratory Syndrome Coronavirus 2. *Emerg. Infect. Dis.* **2020**, *26*, 1654–1665. [CrossRef]
13. Weinberg, G.A.; Schnabel, K.C.; Erdman, D.D.; Prill, M.M.; Iwane, M.K.; Shelley, L.M.; Whitaker, B.L.; Szilagyi, P.G.; Hall, C.B. Field evaluation of TaqMan Array Card (TAC) for the simultaneous detection of multiple respiratory viruses in children with acute respiratory infection. *J. Clin. Virol.* **2013**, *57*, 254–260. [CrossRef] [PubMed]
14. Fry, A.M.; Chittaganpitch, M.; Baggett, H.C.; Peret, T.C.T.; Dare, R.K.; Sawatwong, P.; Thamthitiwat, S.; Areerat, P.; Sanasuttipun, W.; Fischer, J.; et al. The Burden of Hospitalized Lower Respiratory Tract Infection due to Respiratory Syncytial Virus in Rural Thailand. *PLoS ONE* **2010**, *5*, e15098. [CrossRef] [PubMed]
15. Piedimonte, G.; Perez, M.K. Respiratory syncytial virus infection and bronchiolitis. *Pediatr. Rev.* **2014**, *35*, 519–530. [CrossRef] [PubMed]
16. Thorburn, K.; Harigopal, S.; Reddy, V.; Taylor, N.; van Saene, H.K. High incidence of pulmonary bacterial co-infection in children with severe respiratory syncytial virus (RSV) bronchiolitis. *Thorax* **2006**, *61*, 611–615. [CrossRef] [PubMed]
17. Kabir, A.L.; Rahman, A.F.; Rahman, A. ARI situation in our country: Aren't we oblivious of bronchiolitis in Bangladesh? *Mymensingh Med. J.* **2009**, *18*, S50–S55. [PubMed]
18. Stockman, L.J.; Brooks, W.A.; Streatfield, P.K.; Rahman, M.; Goswami, D.; Nahar, K.; Rahman, M.Z.; Luby, S.P.; Anderson, L.J. Challenges to evaluating respiratory syncytial virus mortality in Bangladesh, 2004–2008. *PLoS ONE* **2013**, *8*, e53857. [CrossRef] [PubMed]
19. Liu, P.; Xu, M.; Cao, L.; Su, L.; Lu, L.; Dong, N.; Jia, R.; Zhu, X.; Xu, J. Impact of COVID-19 pandemic on the prevalence of respiratory viruses in children with lower respiratory tract infections in China. *Virol. J.* **2021**, *18*, 159. [CrossRef]
20. Redlberger-Fritz, M.; Kundi, M.; Aberle, S.W.; Puchhammer-Stockl, E. Significant impact of nationwide SARS-CoV-2 lockdown measures on the circulation of other respiratory virus infections in Austria. *J. Clin. Virol.* **2021**, *137*, 104795. [CrossRef]
21. Angoulvant, F.; Ouldali, N.; Yang, D.D.; Filser, M.; Gajdos, V.; Rybak, A.; Guedj, R.; Soussan-Banini, V.; Basmaci, R.; Lefevre-Utile, A.; et al. Coronavirus Disease 2019 Pandemic: Impact Caused by School Closure and National Lockdown on Pediatric Visits and Admissions for Viral and Nonviral Infections-a Time Series Analysis. *Clin. Infect. Dis.* **2021**, *72*, 319–322. [CrossRef]
22. Friedrich, F.; Ongaratto, R.; Scotta, M.C.; Veras, T.N.; Stein, R.T.; Lumertz, M.S.; Jones, M.H.; Comaru, T.; Pinto, L.A. Early Impact of Social Distancing in Response to Coronavirus Disease 2019 on Hospitalizations for Acute Bronchiolitis in Infants in Brazil. *Clin. Infect. Dis.* **2021**, *72*, 2071–2075. [CrossRef]
23. Sherman, A.C.; Babiker, A.; Sieben, A.J.; Pyden, A.; Steinberg, J.; Kraft, C.S.; Koelle, K.; Kanjilal, S. The Effect of Severe Acute Respiratory Syndrome Coronavirus 2 (SARS-CoV-2) Mitigation Strategies on Seasonal Respiratory Viruses: A Tale of 2 Large Metropolitan Centers in the United States. *Clin. Infect. Dis.* **2021**, *72*, e154–e157. [CrossRef] [PubMed]
24. Van Brusselen, D.; De Troeyer, K.; Ter Haar, E.; Vander Auwera, A.; Poschet, K.; Van Nuijs, S.; Bael, A.; Stobbelaar, K.; Verhulst, S.; Van Herendael, B.; et al. Bronchiolitis in COVID-19 times: A nearly absent disease? *Eur. J. Pediatr.* **2021**, *180*, 1969–1973. [CrossRef] [PubMed]
25. Di Mattia, G.; Nenna, R.; Mancino, E.; Rizzo, V.; Pierangeli, A.; Villani, A.; Midulla, F. During the COVID-19 pandemic where has respiratory syncytial virus gone? *Pediatr. Pulmonol.* **2021**, *56*, 3106–3109. [CrossRef] [PubMed]
26. Oh, D.Y.; Buda, S.; Biere, B.; Reiche, J.; Schlosser, F.; Duwe, S.; Wedde, M.; von Kleist, M.; Mielke, M.; Wolff, T.; et al. Trends in respiratory virus circulation following COVID-19-targeted nonpharmaceutical interventions in Germany, January–September 2020: Analysis of national surveillance data. *Lancet Reg. Health Eur.* **2021**, *6*, 100112. [CrossRef] [PubMed]
27. Fourgeaud, J.; Toubiana, J.; Chappuy, H.; Delacourt, C.; Moulin, F.; Parize, P.; Scemla, A.; Abid, H.; Leruez-Ville, M.; Frange, P. Impact of public health measures on the post-COVID-19 respiratory syncytial virus epidemics in France. *Eur. J. Clin. Microbiol. Infect. Dis.* **2021**, *40*, 2389–2395. [CrossRef] [PubMed]
28. Foley, D.A.; Yeoh, D.K.; Minney-Smith, C.A.; Martin, A.C.; Mace, A.O.; Sikazwe, C.T.; Le, H.; Levy, A.; Moore, H.C.; Blyth, C.C. The Interseasonal Resurgence of Respiratory Syncytial Virus in Australian Children Following the Reduction of Coronavirus Disease 2019-Related Public Health Measures. *Clin. Infect. Dis.* **2021**, *73*, e2829–e2830. [CrossRef] [PubMed]
29. Nygaard, U.; Hartling, U.B.; Nielsen, J.; Vestergaard, L.S.; Dungu, K.H.S.; Nielsen, J.S.A.; Sellmer, A.; Matthesen, A.T.; Kristensen, K.; Holm, M. Hospital admissions and need for mechanical ventilation in children with respiratory syncytial virus before and during the COVID-19 pandemic: A Danish nationwide cohort study. *Lancet Child Adolesc. Health* **2023**, *7*, 171–179. [CrossRef]
30. Dolores, A.; Stephanie, G.; Mercedes, S.N.; Erica, G.; Mistchenko, A.S.; Mariana, V. RSV reemergence in Argentina since the SARS-CoV-2 pandemic. *J. Clin. Virol.* **2022**, *149*, 105126. [CrossRef]
31. Mondal, P.; Sinharoy, A.; Gope, S. The Influence of COVID-19 on Influenza and Respiratory Syncytial Virus Activities. *Infect. Dis. Rep.* **2022**, *14*, 134–141. [CrossRef]
32. Ogilvie, M.M.; Vathenen, A.S.; Radford, M.; Codd, J.; Key, S. Maternal antibody and respiratory syncytial virus infection in infancy. *J. Med. Virol.* **1981**, *7*, 263–271. [CrossRef]
33. Piedra, P.A. Clinical experience with respiratory syncytial virus vaccines. *Pediatr. Infect. Dis. J.* **2003**, *22*, S94–S99. [CrossRef] [PubMed]
34. Rodriguez-Fernandez, R.; Mejias, A.; Ramilo, O. Monoclonal Antibodies for Prevention of Respiratory Syncytial Virus Infection. *Pediatr. Infect. Dis. J.* **2021**, *40*, S35–S39. [CrossRef] [PubMed]

35. Balaguer, M.; Alejandre, C.; Vila, D.; Esteban, E.; Carrasco, J.L.; Cambra, F.J.; Jordan, I. Bronchiolitis Score of Sant Joan de Deu: BROSJOD Score, validation and usefulness. *Pediatr. Pulmonol.* **2017**, *52*, 533–539. [CrossRef] [PubMed]
36. Mount, M.C.; Ji, X.; Kattan, M.W.; Slain, K.N.; Clayton, J.A.; Rotta, A.T.; Shein, S.L. Derivation and Validation of the Critical Bronchiolitis Score for the PICU. *Pediatr. Crit. Care Med.* **2022**, *23*, e45–e54. [CrossRef] [PubMed]
37. Hossain, S.J.; Ferdousi, M.J.; Siddique, M.A.B.; Tipu, S.; Qayyum, M.A.; Laskar, M.S. Self-reported health problems, health care seeking behaviour and cost coping mechanism of older people: Implication for primary health care delivery in rural Bangladesh. *J. Fam. Med. Prim. Care* **2019**, *8*, 1209–1215. [CrossRef]

MDPI AG
Grosspeteranlage 5
4052 Basel
Switzerland
Tel.: +41 61 683 77 34

Viruses Editorial Office
E-mail: viruses@mdpi.com
www.mdpi.com/journal/viruses

www.ingramcontent.com/pod-product-compliance
Lightning Source LLC
Chambersburg PA
CBHW041342210926
51850CB00013B/2952